Student Solutions Manual

for use with

Elementary Statistics
A Step by Step Approach

Tenth Edition

Allan G. Bluman

McGraw Hill Education

STUDENT SOLUTIONS MANUAL FOR USE WITH ELEMENTARY STATISTICS: A STEP BY STEP APPROACH, TENTH EDITION

Published by McGraw-Hill Education, 2 Penn Plaza, New York, NY 10121. Copyright © 2018 by McGraw-Hill Education. All rights reserved. Printed in the United States of America. Previous editions © 2010, 2012, and 2014. No part of this publication may be reproduced or distributed in any form or by any means, or stored in a database or retrieval system, without the prior written consent of McGraw-Hill Education, including, but not limited to, in any network or other electronic storage or transmission, or broadcast for distance learning.

Some ancillaries, including electronic and print components, may not be available to customers outside the United States.

This book is printed on acid-free paper.

1 2 3 4 5 6 QVS 21 20 19 18 17

ISBN 978-1-260-04206-1
MHID 1-260-04206-5

All credits appearing on page or at the end of the book are considered to be an extension of the copyright page.

The Internet addresses listed in the text were accurate at the time of publication. The inclusion of a website does not indicate an endorsement by the authors or McGraw-Hill Education, and McGraw-Hill Education does not guarantee the accuracy of the information presented at these sites.

mheducation.com/highered

Student Solution Manual

Elementary Statistics
Tenth Edition

Table of Contents

Chapter 1: The Nature of Probability and Statistics 1

Chapter 2: Frequency Distributions and Graphs 5

Chapter 3: Data Description 19

Chapter 4: Probability and Counting Rules 32

Chapter 5: Discrete Probability Distributions 47

Chapter 6: The Normal Distribution 58

Chapter 7: Confidence Intervals and Sample Size 75

Chapter 8: Hypothesis Testing 82

Chapter 9: Testing the Difference Between Two Means, Two Proportions, and Two Variances 98

Chapter 10: Correlation and Regression 115

Chapter 11: Other Chi-Square Tests 127

Chapter 12: Analysis of Variance 138

Chapter 13: Nonparametric Statistics 149

Chapter 14: Sampling and Simulation 165

Chapter 1 - The Nature of Probability and Statistics

EXERCISE SET 1-1

1. Statistics is the science of conducting studies to collect, organize, summarize, analyze, and draw conclusions from data.

3. In a census, the researchers collect data from all subjects in the population.

5. Descriptive statistics consists of the collection, organization, summarization, and presentation of data while inferential statistics consists of generalizing from samples to populations, performing estimations and hypothesis testing, determining relationships among variables, and making predictions.

7. Samples are used more than populations both because populations are usually large and because researchers are unable to use every subject in the population.

9. This is inferential because a generalization is being made about the population.

11. This is a descriptive statistic since it describes the weight loss for a specific group of subjects, i.e., the teenagers at Boston University.

13. This is an inferential statistic since a generalization has been made about the population.

15. This is an inferential statistic since a generalization was made about the population.

17. This is an inferential statistic since it is a generalization made from data obtained from a sample.

19. Answers will vary.

EXERCISE SET 1-2

1. Qualitative variables are variables that can be placed in distinct categories according to some characteristic or attribute and cannot be ranked; while quantitative variables are numerical in nature and can be ordered or counted.

3. Continuous variables need to be rounded because of the limits of the measuring device.

5. Qualitative 7. Quantitative

9. Quantitative 11. Discrete

13. Continuous 15. Discrete

17. 23.5-24.5 feet

19. 142.5-143.5 miles

21. 200.65-200.75 miles

23. Nominal 25. Ratio

27. Ordinal 29. Ratio

EXERCISE SET 1-3

1. Data can be collected by using telephone surveys, mail questionnaire surveys, personal interview surveys, by taking a look at records, or by direct observation methods.

3. Random numbers are used in sampling so that every subject in the population has an equal chance of being selected for a sample. Random numbers can be generated by computers or calculators; however, there are other ways of generating random numbers such as using a random number table or rolling dice.

Chapter 1 - The Nature of Probability and Statistics

5. The population could be all people in the United States who earn over $200,000 per annum. A sample could have been created by selecting 500 people randomly from an accounting firm that prepares income taxes. Answers will vary.

7. The population could be all households in the United States. A sample could be selected using 1000 households in the United States. Answers will vary.

9. The population could be all adults in the United States who develop diabetes. The sample could be surveying patient records of these people to see if they have been taking statins. Again, the privacy rights must be considered. Answers will vary.

11. Systematic   13. Random

15. Cluster

EXERCISE SET 1-4

1. In an observational study, the researcher observes what is happening and tries to draw conclusions based on the observations. In an experimental study, the researcher manipulates one of the variables and tries to determine how this influences the variables.

3. One advantage of an observational study is that it can occur in a natural setting. In addition, researchers can look at past instances of statistics and draw conclusions from these situations. Another advantage is that the researcher can use variables, such as drugs, that he or she cannot manipulate.

3. continued
One disadvantage is that since the variable cannot be manipulated, a definite cause-and-effect situation cannot be shown. Another disadvantage is that these studies can be expensive and time-consuming. These studies can also be influenced by confounding variables. Finally, in these studies, the researcher sometimes needs to rely on data collected by others.

5. In an experimental study, the researcher has control of the assignment of subjects to the groups whereas in a quasi-experimental study, the researcher uses intact groups.

7. In research studies, a treatment group subject receives a specific treatment while those in the control group do not receive a treatment or are given a placebo.

9. A confounding variable is one that can influence the results of the research study when no precautions were taken to eliminate it from the study.

11. Blinding is used to help eliminate the placebo effect. Here the subjects are given a sugar pill that looks like the real medical pill. The subjects do not know which pill they are getting. When double blinding occurs, neither the subjects nor the researchers are told who gets the real treatment or the placebo.

13. In a completely randomized design, the subjects are assigned to the groups randomly, whereas in a matched-pair design, subjects are matched on some variable.

Chapter 1 - The Nature of Probability and Statistics

13. continued

Then one subject is randomly assigned to one group, and the other subject is assigned to the other group. In both types of studies, the treatments can be randomly assigned to the groups.

15. Observational

17. Experimental

19. Independent variable - minutes exercising Dependent variable - catching a cold

21. Independent variable - happy face on the check
Dependent variable - amount of the tip

23. Age, income, socioeconomic status. Answers will vary.

25. Income, number of hours worked, type of boss. Answers will vary.

27. How is a perfect body defined statistically?

29. How can 24 hours of pain relief be measured?

31. How much weight, if any, will be lost?

33. Only 20 people were used in the study.

35. It is meaningless since there is no definition of "the road less traveled." Also, there is no way to know that for every 100 women, 91 would say that they have taken "the road less traveled."

37. There is no mention of how this conclusion was obtained.

39. Since the word may is used, there is no guarantee that the product will help fight cancer and heart disease.

41. No. There are many other factors that contribute to criminal behavior.

43. Answers will vary.

45. Answers will vary.

REVIEW EXERCISES - CHAPTER 1

1. Inferential   3. Descriptive

5. Inferential   7. Descriptive

9. Ratio   11. Interval

13. Ratio   15. Ordinal

17. Ratio   19. Qualitative

21. Quantitative   23. Quantitative

25. Quantitative   27. Discrete

29. Discrete   31. Continuous

33. Continuous   35. 55.5-56.5 yards

37. 72.55-72.65 tons

39. Cluster

41. Random

43. Stratified

45. Experimental

47. Observational

Chapter 1 - The Nature of Probability and Statistics

49. Independent variable—habitat of the animal
Dependent variable—weight of the animal

51. Independent variable - thyme
Dependent variable - antioxidants

53. A telephone survey won't contact all the types of people who shop online. Answers will vary.

55. It depends on where the survey was conducted. Some places in the United States receive very little or no snowfall at all during winter.

57. How can the Internet raise IQ? Answers will vary.

CHAPTER QUIZ

1. True

3. False, nonsampling error is the result of collecting data incorrectly or selecting a biased sample.

5. True

7. False, it is 5.5-6.5 inches.

9. b

11. a   13. a

15. Gambling, insurance. Answers can vary.

17. Sample

19.
 a. Random   c. Cluster
 b. Systematic d. Stratified

21. Random

23.
 a. Nominal   d. Interval
 b. Ratio   e. Ratio
 c. Ordinal

25.
 a. 31.5 – 32.5 minutes
 b. 0.475 – 0.485 millimeter
 c. 6.15 – 6.25 inches
 d. 18.5 – 19.5 pounds
 e. 12.05 – 12.15 quarts

Chapter 2 - Frequency Distributions and Graphs

EXERCISE SET 2-1

1. Frequency distributions are used to organize data in a meaningful way, to determine the shape of the distribution, to facilitate computational procedures for statistics, to make it easier to draw charts and graphs, and to make comparisons among different sets of data.

3. Five to twenty classes. Width should be an odd number so that the midpoint will have the same place value as the data.

5.
Boundaries: 57.5 – 62.5
Midpoint: 60
Width: 5

7.
Boundaries: 16.345 – 18.465
Midpoint: 17.405
Width: 2.12

9. Class width is not uniform.

11. A class has been omitted.

13.

Class	f	Percent
V	6	12
C	7	14
M	22	44
H	3	6
P	12	24
	50	100

The mocha flavor class has the most data values and the hazelnut class has the least number of data values.

15.

Limits	Boundaries	f
0	-0.5 - 0.5	2
1	0.5 - 1.5	5
2	1.5 - 2.5	24
3	2.5 - 3.5	8
4	3.5 - 4.5	6
5	4.5 - 5.5	4
6	5.5 - 6.5	0
7	6.5 - 7.5	1
		50

	cf
Less than -0.5	0
Less than 0.5	2
Less than 1.5	7
Less than 2.5	31
Less than 3.5	39
Less than 4.5	45
Less than 5.5	49
Less than 6.5	49
Less than 7.5	50

The category "twice a week" has more values than any other category.

17.
H = 93   L = 48
Range = 93 − 48 = 45
Width = 45 ÷ 7 = 6.4 round up to 7

Limits	Boundaries	f
48 - 54	47.5 - 54.5	3
55 - 61	54.5 - 61.5	2
62 - 68	61.5 - 68.5	9
69 - 75	68.5 - 75.5	13
76 - 82	75.5 - 82.5	8
83 - 89	82.5 - 89.5	3
90 - 96	89.5 - 96.5	2
		40

Chapter 2 - Frequency Distributions and Graphs

17. continued

	cf
Less than 47.5	0
Less than 54.5	3
Less than 61.5	5
Less than 68.5	14
Less than 75.5	27
Less than 82.5	35
Less than 89.5	38
Less than 96.5	40

19.
H = 70   L = 27
Range = 70 − 27 = 43
Width = 43 ÷ 7 = 6.1 or 7

Limits	Boundaries	f
27 - 33	26.5 - 33.5	7
34 - 40	33.5 - 40.5	14
41 - 47	40.5 - 47.5	15
48 - 54	47.5 - 54.5	11
55 - 61	54.5 - 61.5	3
62 - 68	61.5 - 68.5	3
69 - 75	68.5 - 75.5	2
		55

	cf
Less than 26.5	0
Less than 33.5	7
Less than 40.5	21
Less than 47.5	36
Less than 54.5	47
Less than 61.5	50
Less than 68.5	53
Less than 75.5	55

21.
H = 88   L = 12
Range = 88 − 12 = 76
Width = 76 ÷ 9 = 8.4 round up to 9

21. continued

Limits	Boundaries	f
12 - 20	11.5 - 20.5	7
21 - 29	20.5 - 29.5	7
30 - 38	29.5 - 38.5	3
39 - 47	38.5 - 47.5	3
48 - 56	47.5 - 56.5	4
57 - 65	56.5 - 65.5	3
66 - 74	65.5 - 74.5	0
75 - 83	74.5 - 83.5	2
84 - 92	83.5 - 92.5	1
		30

	cf
Less than 11.5	0
Less than 20.5	7
Less than 29.5	14
Less than 38.5	17
Less than 47.5	20
Less than 56.5	24
Less than 65.5	27
Less than 74.5	27
Less than 83.5	29
Less than 92.5	30

23.
H = 49   L = 14
Range = 49 − 14 = 35
Width = 7

Limits	Boundaries	f
14 - 20	13.5 - 20.5	10
21 - 27	20.5 - 27.5	11
28 - 34	27.5 - 34.5	6
35 - 41	34.5 - 41.5	8
42 - 48	41.5 - 48.5	4
49 - 55	48.5 - 55.5	1
		40

Chapter 2 - Frequency Distributions and Graphs

23. continued

	cf
Less than 13.5	0
Less than 20.5	10
Less than 27.5	21
Less than 34.5	27
Less than 41.5	35
Less than 48.5	39
Less than 55.5	40

25.
H = 12.3   L = 6.2
Range = 12.3 − 6.2 = 6.1
Width = 6.1 ÷ 7 = 0.87 round up to 0.9

Limits	Boundaries	f
6.2 - 7.0	6.15 - 7.05	1
7.1 - 7.9	7.05 - 7.95	7
8.0 - 8.8	7.95 - 8.85	9
8.9 - 9.7	8.85 - 9.75	7
9.8 - 10.6	9.75 - 10.65	8
10.7 - 11.5	10.65 - 11.55	4
11.6 - 12.4	11.55 - 12.45	4
		40

	cf
Less than 6.15	0
Less than 7.05	1
Less than 7.95	8
Less than 8.85	17
Less than 9.75	24
Less than 10.65	32
Less than 11.55	36
Less than 12.45	40

27. The percents add up to 101%. They should total 100% unless rounding was used.

EXERCISE SET 2-2

1.

Limits	Boundaries	X_m	f
90 - 98	89.5 - 98.5	94	6
99 - 107	98.5 - 107.5	103	22
108 - 116	107.5 - 116.5	112	43
117 - 125	116.5 - 125.5	121	28
126 - 134	125.5 - 134.5	130	9
			108

	cf
Less than 89.5	0
Less than 98.5	6
Less than 107.5	28
Less than 116.5	71
Less than 125.5	99
Less than 134.5	108

Eighty applicants do not need to enroll in the developmental programs.

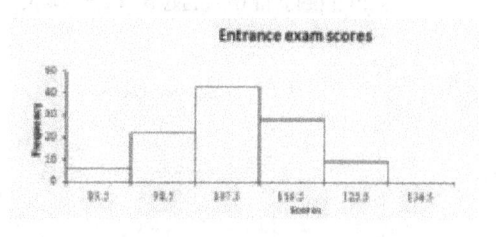

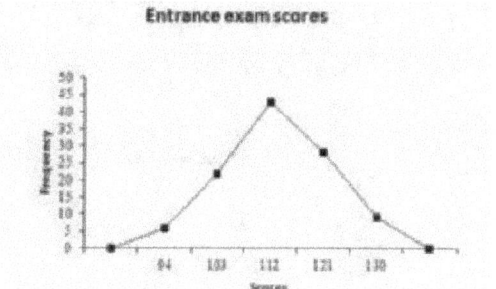

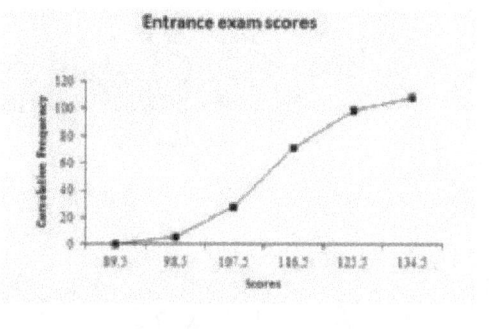

Chapter 2 - Frequency Distributions and Graphs

3.

Limits	Boundaries	X_m	f
9 - 11	8.5 - 11.5	10	2
12 - 14	11.5 - 14.5	13	20
15 - 17	14.5 - 17.5	16	18
18 - 20	17.5 - 20.5	19	7
21 - 23	20.5 - 23.5	22	2
24 - 26	23.5 - 26.5	25	1
			50

	cf
Less than 8.5	0
Less than 11.5	2
Less than 14.5	22
Less than 17.5	40
Less than 20.5	47
Less than 23.5	49
Less than 26.5	50

The distribution is positively skewed with a peak at the class of 11.5–14.5.

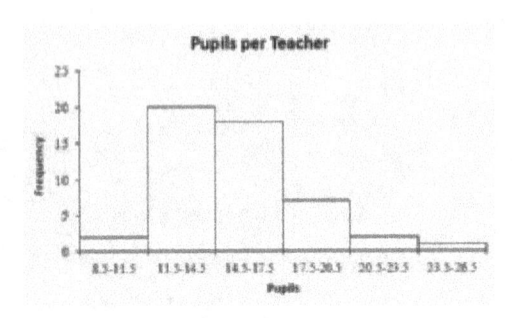

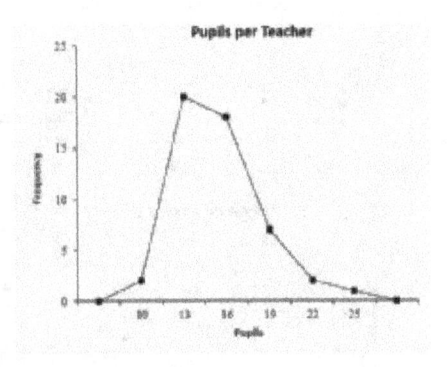

3. continued

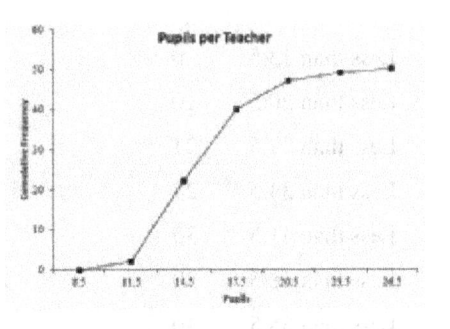

5.

Limits	Boundaries	X_m	f
1 - 43	0.5 - 43.5	22	24
44 - 86	43.5 - 86.5	65	17
87 - 129	86.5 - 129.5	108	3
130 - 172	129.5 - 172.5	151	4
173 - 215	172.5 - 215.5	194	1
216 - 258	215.5 - 258.5	237	0
259 - 301	258.5 - 301.5	280	0
302 - 344	301.5 - 344.5	323	1
			50

	cf
Less than 0.5	0
Less than 43.5	24
Less than 86.5	41
Less than 129.5	44
Less than 172.5	48
Less than 215.5	49
Less than 258.5	49
Less than 301.5	49
Less than 344.5	50

The distribution is positively skewed.

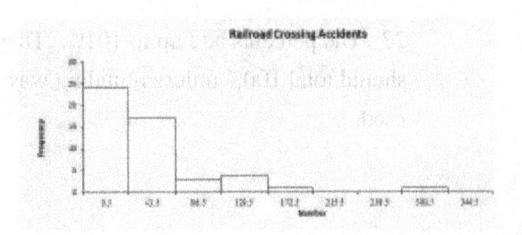

Chapter 2 - Frequency Distributions and Graphs

5. continued

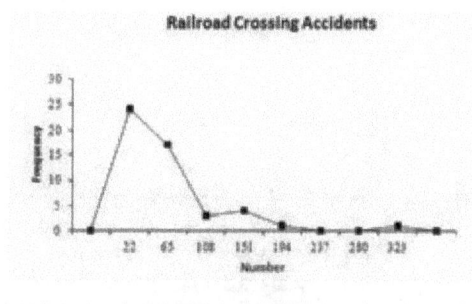

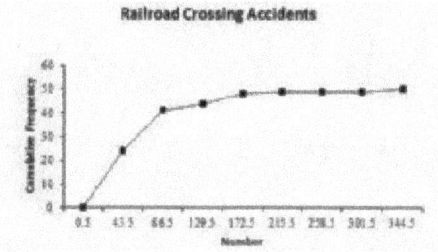

Railroad Crossing Accidents (frequency polygon)

Railroad Crossing Accidents (ogive)

7.

Limits	Boundaries	X_m	f
1260 - 1734	1259.5 - 1734.5	1497	12
1735 - 2209	1734.5 - 2209.5	1972	6
2210 - 2684	2209.5 - 2684.5	2447	3
2685 - 3159	2684.5 - 3159.5	2922	1
3160 - 3634	3159.5 - 3634.5	3397	1
3635 - 4109	3634.5 - 4109.5	3872	1
4110 - 4584	4109.5 - 4584.5	4347	2
			26

	cf
Less than 1259.5	0
Less than 1734.5	12
Less than 2209.5	18
Less than 2684.5	21
Less than 3159.5	22
Less than 3634.5	23
Less than 4109.5	24
Less than 4584.5	26

The distribution is positively skewed. The class with the most frequencies is 1259.5 - 1734.5.

7. continued

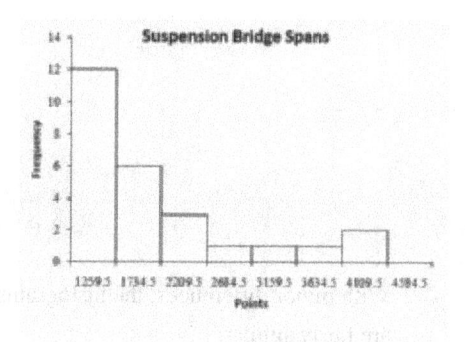

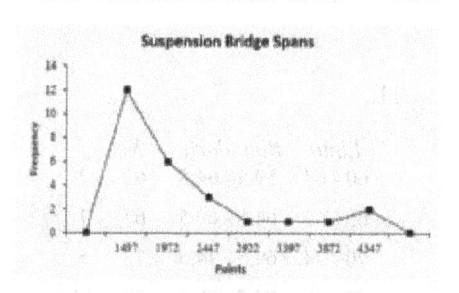

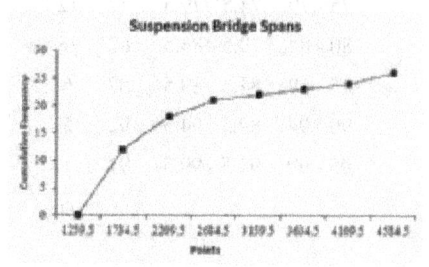

Suspension Bridge Spans (histogram)

Suspension Bridge Spans (frequency polygon)

Suspension Bridge Spans (ogive)

9.

Limits	Boundaries	f(now)	f(5 years ago)
10 - 14	9.5 - 14.5	6	5
15 - 19	14.5 - 19.5	4	4
20 - 24	19.5 - 24.5	3	2
25 - 29	24.5 - 29.5	2	3
30 - 34	29.5 - 34.5	5	6
35 - 39	34.5 - 39.5	1	2
40 - 44	39.5 - 44.5	2	1
45 - 49	44.5 - 49.5	1	1
Total		24	24

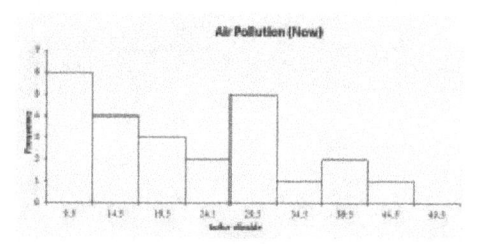

Air Pollution (Now)

Chapter 2 - Frequency Distributions and Graphs

9. continued

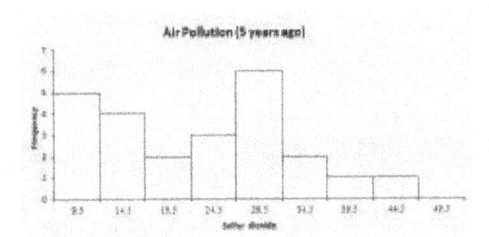

With minor differences, the histograms are fairly similar.

11.

Limits	Boundaries	X_m	f
60 - 64	59.5 - 64.5	62	2
65 - 69	64.5 - 69.5	67	1
70 - 74	69.5 - 74.5	72	5
75 - 79	74.5 - 79.5	77	12
80 - 84	79.5 - 84.5	82	18
85 - 89	84.5 - 89.5	87	6
90 - 94	89.5 - 94.5	92	5
95 - 99	94.5 - 99.5	97	<u>1</u>
			50

	cf
Less than 59.5	0
Less than 64.5	2
Less than 69.5	3
Less than 74.5	8
Less than 79.5	20
Less than 84.5	38
Less than 89.5	44
Less than 94.5	49
Less than 99.5	50

Most patients fell into the 75–84 range.

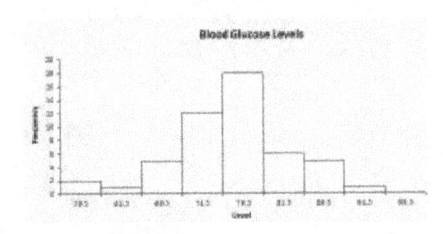

11. continued

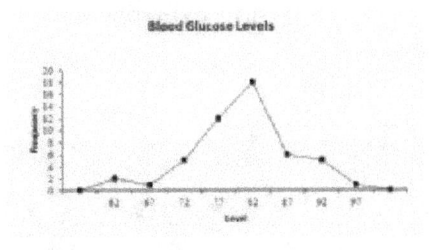

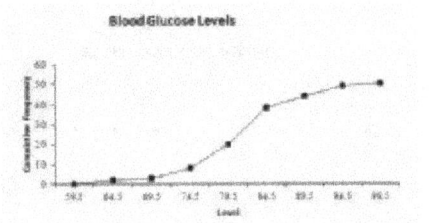

13.

Boundaries	X_m	rf
89.5 - 98.5	94	0.06
98.5 - 107.5	103	0.20
107.5 - 116.5	112	0.40
116.5 - 125.5	121	0.26
125.5 - 134.5	130	<u>0.08</u>
		1.00

	crf
Less than 89.5	0
Less than 98.5	0.06
Less than 107.5	0.26
Less than 116.5	0.66
Less than 125.5	0.92
Less than 134.5	1.00

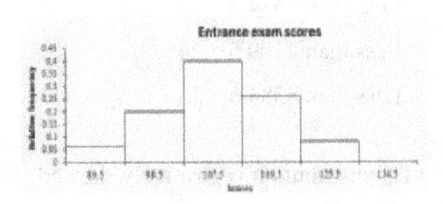

Chapter 2 - Frequency Distributions and Graphs

13. continued

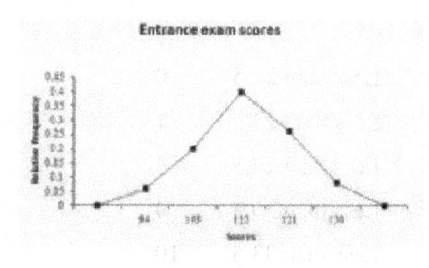

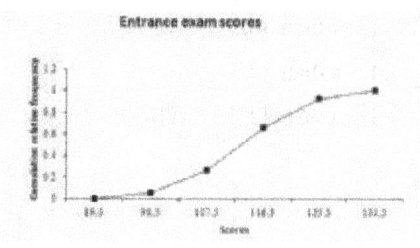

The proportion of applicants who do not need to enroll in the development program is about 0.74.

15.

Boundaries	X_m	rf
0.5 - 43.5	22	0.48
43.5 - 86.5	65	0.34
86.5 - 129.5	108	0.06
129.5 - 172.5	151	0.08
172.5 - 215.5	194	0.02
215.5 - 258.5	237	0.00
258.5 - 301.5	280	0.00
301.5 - 344.5	323	0.02
		1.00

	crf
Less than 0.5	0
Less than 43.5	0.48
Less than 86.5	0.82
Less than 129.5	0.88
Less than 172.5	0.96
Less than 215.5	0.98
Less than 258.5	0.98
Less than 301.5	0.98
Less than 344.5	1.00

15. continued

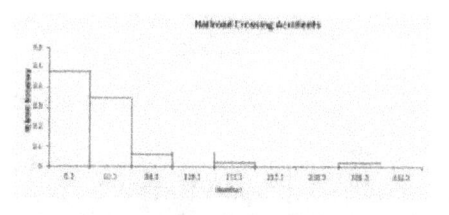

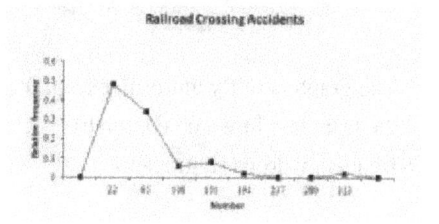

17.

Boundaries	X_m	rf
35.5 - 40.5	38	0.23
40.5 - 45.5	43	0.20
45.5 - 50.5	48	0.23
50.5 - 55.5	53	0.23
55.5 - 60.5	58	0.10
		0.99*

*due to rounding

	crf
Less than 35.5	0.00
Less than 40.5	0.23
Less than 45.5	0.43
Less than 50.5	0.66
Less than 55.5	0.89
Less than 60.5	0.99

Chapter 2 - Frequency Distributions and Graphs

17. continued

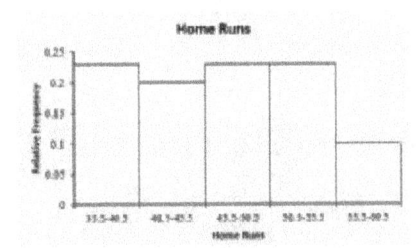

The graph is fairly uniform, except for the last class in which the relative frequency drops significantly.

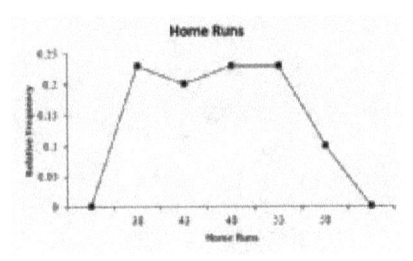

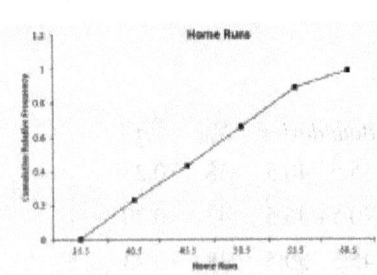

19.

Limits	Boundaries	X_m	f
22 - 24	21.5 - 24.5	23	1
25 - 27	24.5 - 27.5	26	3
28 - 30	27.5 - 30.5	29	0
31 - 33	30.5 - 33.5	32	6
34 - 36	33.5 - 36.5	35	5
37 - 39	36.5 - 39.5	38	3
40 - 42	39.5 - 42.5	41	2
			20

19. continued

	cf
Less than 21.5	0
Less than 24.5	1
Less than 27.5	4
Less than 30.5	4
Less than 33.5	10
Less than 36.5	15
Less than 39.5	18
Less than 42.5	20

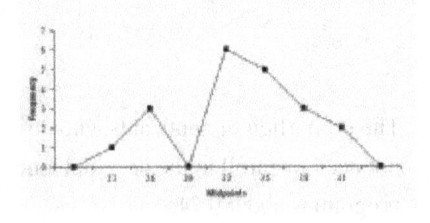

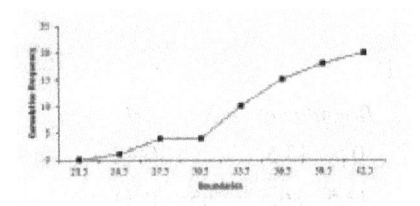

21.

Boundaries	X_m	f
468.5 - 495.5	482	6
495.5 - 522.5	509	15
522.5 - 549.5	536	10
549.5 - 576.5	563	7
576.5 - 603.5	590	6
603.5 - 630.5	617	6
		50

	cf
Less than 468.5	0
Less than 495.5	6
Less than 522.5	21
Less than 549.5	31
Less than 576.5	38
Less than 603.5	44
Less than 630.5	50

Chapter 2 - Frequency Distributions and Graphs

21. continued

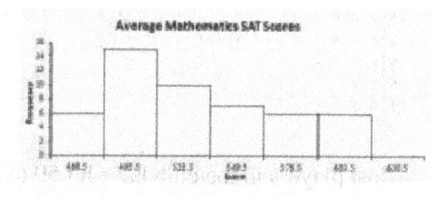

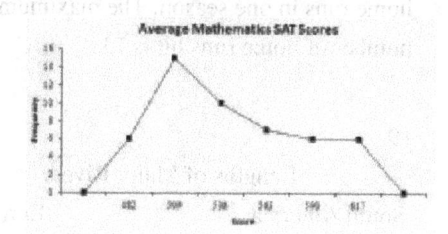

[Histogram: Average Mathematics SAT Scores]

[Frequency polygon: Average Mathematics SAT Scores]

EXERCISE SET 2-3

1.

	f
IBM	380
Hewlett Packard	302
Xerox	147
Microsoft	128
Intel	107

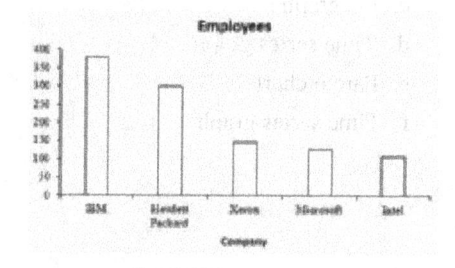

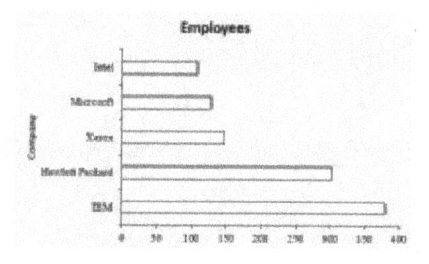

[Vertical bar graph: Employees by Company]

[Horizontal bar graph: Employees by Company]

3.

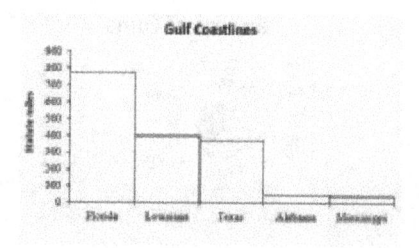

[Bar graph: Gulf Coastlines — Florida, Louisiana, Texas, Alabama, Mississippi]

5.

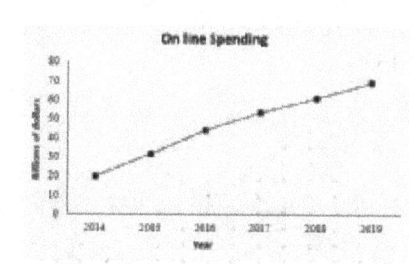

[Line graph: On line Spending, 2014–2019]

There is a steady increase over the years.

7.

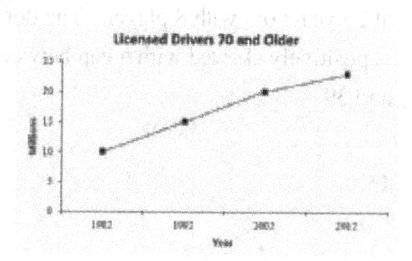

[Line graph: Licensed Drivers 70 and Older, 1982–2012]

9.

[Pie chart: Number of Credit Cards — 0: 26%, 1: 20%, 2 or 3: 34%, 4 or more: 20%]

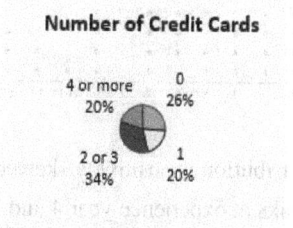

More people have 2 or 3 credit cards.

Chapter 2 - Frequency Distributions and Graphs

11.

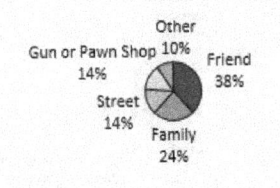

Source of Guns
- Other 10%
- Gun or Pawn Shop 14%
- Friend 38%
- Street 14%
- Family 24%

Guns from friends accounted for 38% of the total usage.

13.

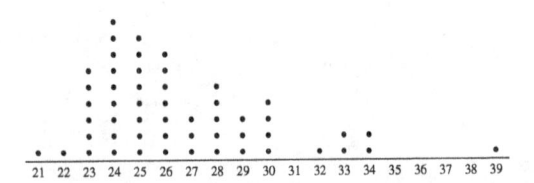

(dotplot with x-axis 21 to 39)

The dotplot is somewhat positively skewed and shows that the majority of the players are between 21 and 30 years old. There are 2 peaks at 24 years old with 9 players, and at 25 years old with 8 players. The dot plot is positively skewed with a gap between 34 and 39.

15.

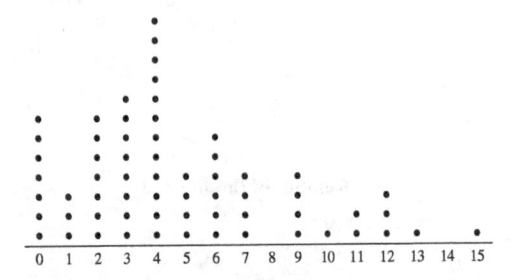

(dotplot with x-axis 0 to 15)

The distribution is positively skewed. The data peaks at experience year 4 and gaps between the experience years of 7 to 9 and 13 to 15. The data clusters between years 0 to 7 and 9 to 13 with a peak at 25 minutes.

17.

```
5 | 0 0 0 0 0 0 1 1 1 1 2 2 2 2 2 3 4 4 4 4 4 4
5 | 6 6 6 7 7 8 8 8 8 9
6 | 0 1 3 4
6 | 5 6
7 | 0 3
```

Most players in the club have hit 50 to 54 home runs in one season. The maximum number of home runs hit is 73.

19.

Lengths of Major Rivers

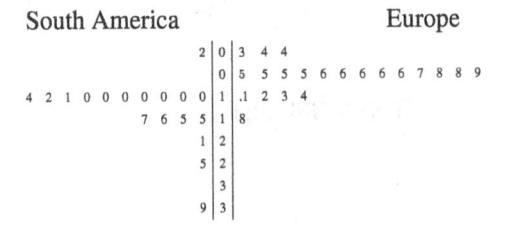

South America   Europe

```
              2 | 0 | 3 4 4
                | 0 | 5 5 5 5 6 6 6 6 7 8 8 9
4 2 1 0 0 0 0 0 0 0 | 1 | 1 2 3 4
        7 6 5 5 | 1 | 8
                | 2 |
              5 | 2 |
                | 3 |
              9 | 3 |
```

The majority of the South American rivers are longer than those in Europe.

21.
a. Pareto chart
b. Pareto chart
c. Pie graph
d. Time series graph
e. Pareto chart
f. Time series graph

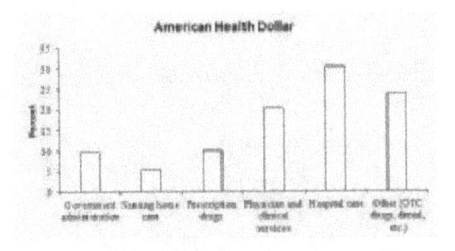

23.

(bar chart titled "American Health Dollar")

Chapter 2 - Frequency Distributions and Graphs

23. continued

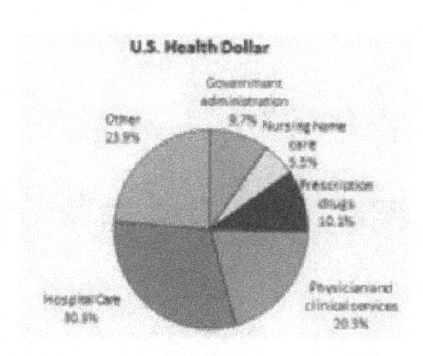

25. The bottle for 2011 is much bigger in area than the bottle for 1988. This causes the eye to see a much bigger difference than the actual difference.

27.

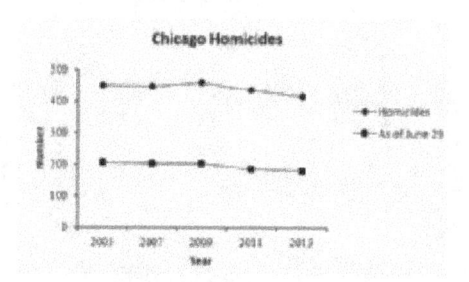

There's no way to tell if the crime rate is decreasing by looking at the graph.

REVIEW EXERCISES - CHAPTER 2

1.

Class	f	Percent
Newspaper	10	20
Television	16	32
Radio	12	24
Internet	12	24
	50	100

3.

Class	f
11	1
12	2
13	2
14	2
15	1
16	2
17	4
18	2
19	2
20	1
21	0
22	1
	20

	cf
less than 10.5	0
less than 11.5	1
less than 12.5	3
less than 13.5	5
less than 14.5	7
less than 15.5	8
less than 16.5	10
less than 17.5	14
less than 18.5	16
less than 19.5	18
less than 20.5	19
less than 21.5	19
less than 22.5	20

5.

Limits	Boundaries	f
53 - 185	52.5 - 185.5	8
186 - 318	185.5 - 318.5	11
319 - 451	318.5 - 451.5	2
452 - 584	451.5 - 584.5	1
585 - 717	584.5 - 717.5	4
718 - 850	717.5 - 850.5	2
		28

Chapter 2 - Frequency Distributions and Graphs

5. continued

	cf
Less than 52.5	0
Less than 185.5	8
Less than 318.5	19
Less than 451.5	21
Less than 584.5	22
Less than 717.5	26
Less than 850.5	28

7.

Limits	Boundaries	rf
53 - 185	52.5 - 185.5	0.29
186 - 318	185.5 - 318.5	0.39
319 - 451	318.5 - 451.5	0.07
452 - 584	451.5 - 584.5	0.04
585 - 717	584.5 - 717.5	0.14
718 - 850	717.5 - 850.5	0.07
		1.00

	crf
Less than 52.5	0
Less than 185.5	0.29
Less than 318.5	0.68
Less than 451.5	0.75
Less than 584.5	0.79
Less than 717.5	0.93
Less than 850.5	1.00

9.

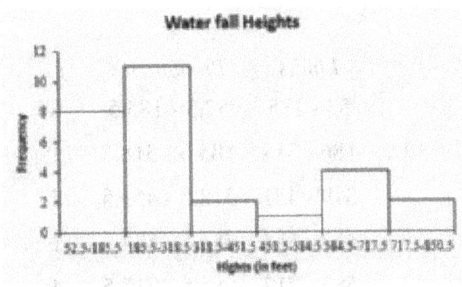

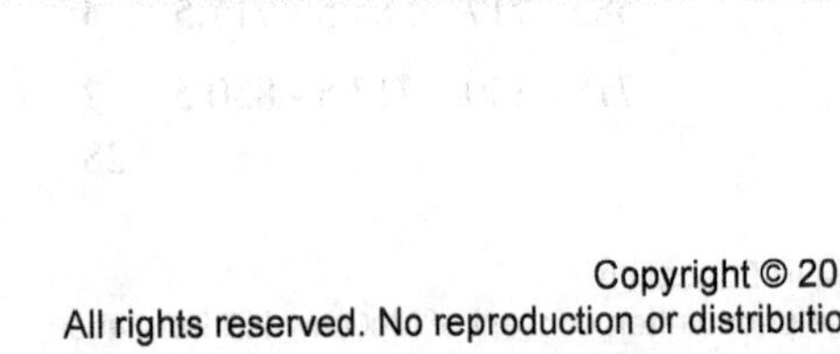

Water fall Heights (histogram)

Water fall Heights (frequency polygon)

Water fall Heights (ogive / cumulative frequency)

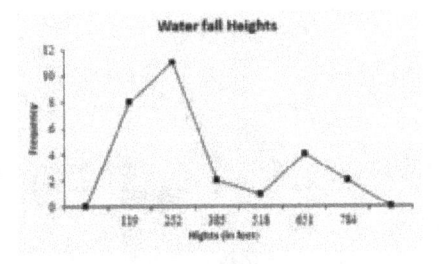

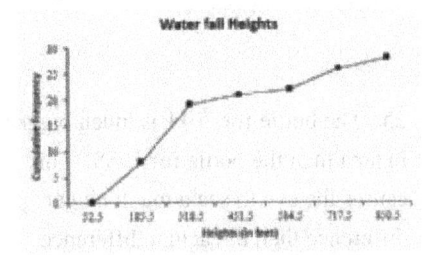

11.

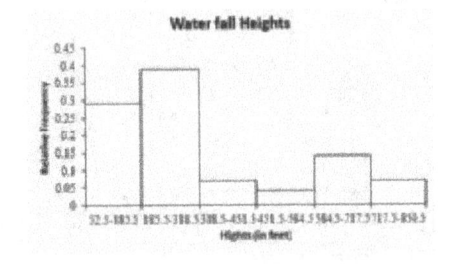

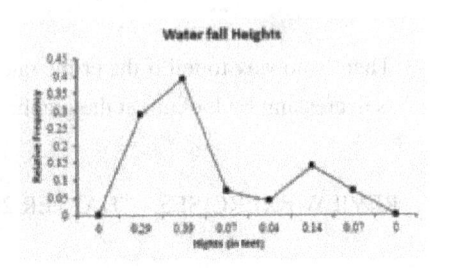

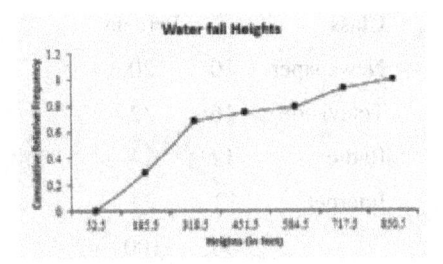

Water fall Heights (relative frequency histogram)

Water fall Heights (relative frequency polygon)

Water fall Heights (cumulative relative frequency ogive)

Chapter 2 - Frequency Distributions and Graphs

13.

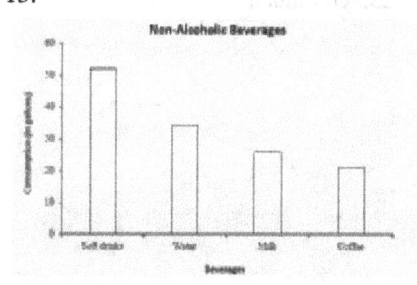

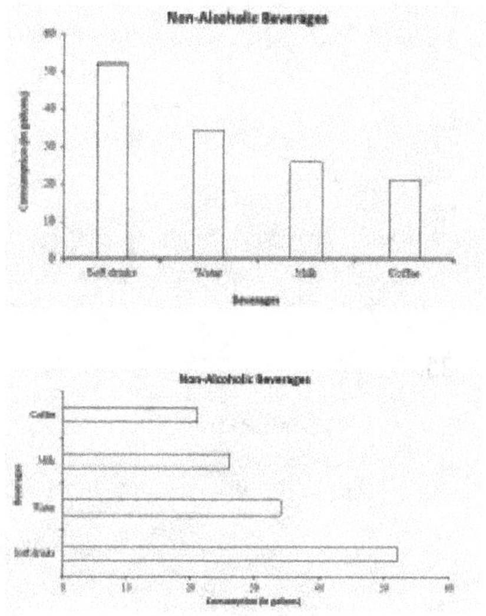

15.

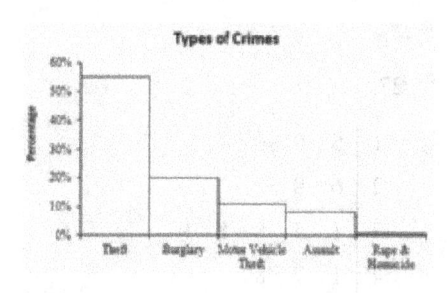

17.

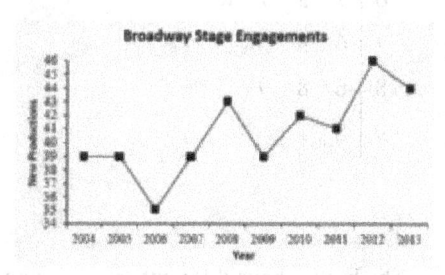

New Productions declined from 2005 to 2006; then, it increased each year until 2008. There was a slight increase in 2010 and 2012.

19.

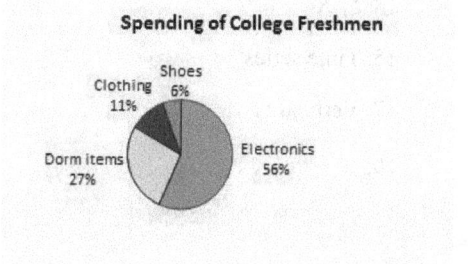

21.

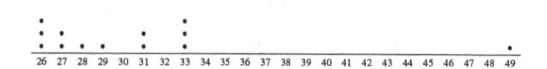

The graph shows almost all but one of the touchdowns per season for Manning's career were between 26 and 33.

23.

20	2	3	6			
21	3	5	8	9	9	
22	0	1	3	3	4	7
23	0	2	3	3	5	8	9
24	6	8	9			
25	4	4	6	8		
26	2	3				

25.
The graphs are misleading because no scale is used on the x and y axes. So it is impossible to tell the times of the pain relief.

CHAPTER 2 QUIZ

1. False

3. False

5. True

7. False

9. c

11. b

Chapter 2 - Frequency Distributions and Graphs

13. 5, 20

15. Time series

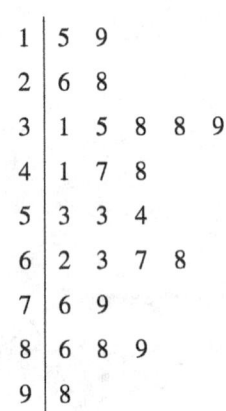

17. Vertical or y

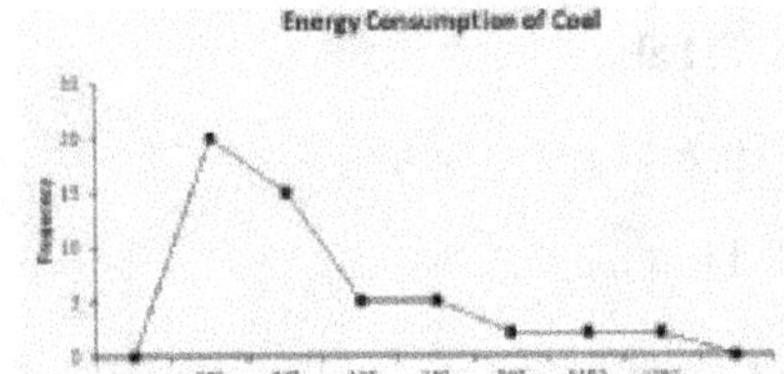

19. [Pie chart: Housing Arrangements — Condominium 32%, House 24%, Apartment 20%, Mobile Home 24%]

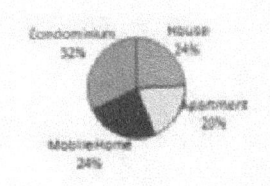

21. [Histogram, frequency polygon, and ogive of Items purchased]

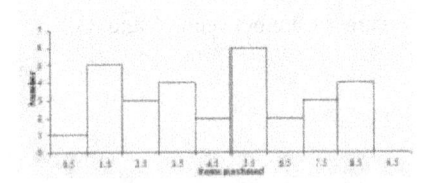

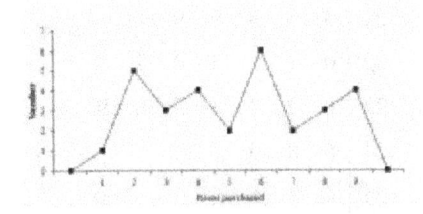

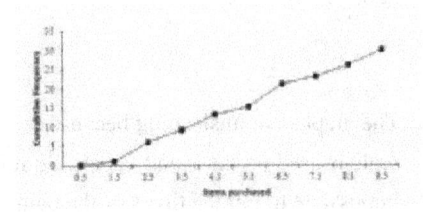

23. [Histogram and frequency polygon: Energy Consumption of Coal]

23. continued [Ogive: Energy Consumption of Coal]

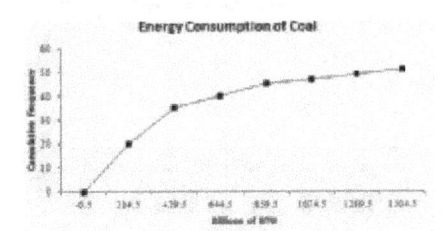

25. [Pie chart: Identify theft — Lost or stolen item 43%, Retail purchases 18%, Stolen mail 11%, Computer hackers 3%, Phishing 5%, Other 12%]

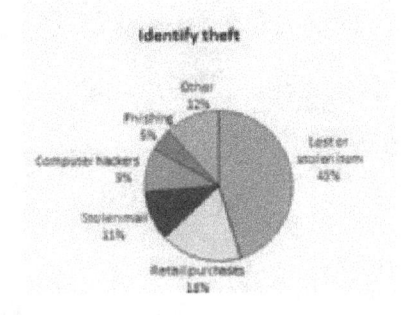

27.
1	5	9		
2	6	8		
3	1	5	8	8	9
4	1	7	8	
5	3	3	4	
6	2	3	7	8
7	6	9		
8	6	8	9	
9	8			

29. The bottles have different areas, so your eyes will compare areas instead of heights.

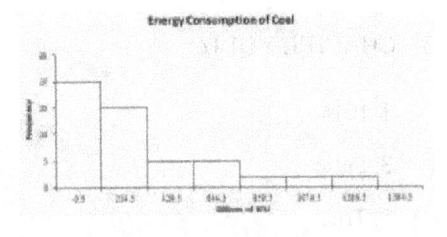

Chapter 3 - Data Description

Note: Answers may vary due to rounding, TI 83's, or computer programs.

EXERCISE SET 3-1

1.
a. $\overline{X} = \frac{\Sigma X}{n} = \frac{1457}{14} = 104.1$

b. MD: 102

c. MR: $\frac{160+50}{2} = 105$

d. Mode: 50, 95, 102, 160

3.
a. $\overline{X} = \frac{\Sigma X}{n} = \frac{1312}{6} = 218.7$

b. MD: 221

c. MR $= \frac{180+251}{2} = 215.5$

d. Mode: no mode

5.
a. $\overline{X} = \frac{\Sigma X}{n} = \frac{10,671,300}{10} = 1,067,130$

b. MD $= \frac{1,100,000+1,210,000}{2} = 1,155,000$

c. Mode: 1,340,000

d. MR $= \frac{298,000+2,000,000}{2} = 1,149,000$

7.
a. $\overline{X} = \frac{\Sigma X}{n} = \frac{289}{11} = 26.3$

b. MD $= 28$

c. MR $= \frac{10+38}{2} = 24$

d. Mode: 30

The mean, median, and midrange are all very close.

9.
a. $\overline{X} = \frac{\Sigma X}{n} = \frac{398.2}{13} = 30.6$

b. MD $= 10$

c. MR $= \frac{3.1+143.8}{2} = 73.45$

d. Mode: no mode

11.
a. $\overline{X} = \frac{\Sigma X}{n} = \frac{486.2}{13} = 37.4$

b. MD $= 33.7$

c. Mode: no mode

d. MR $= \frac{4.4+87.9}{2} = 46.15$

13.

Boundaries	X_m	f	$f \cdot X_m$
47.5 - 54.5	51	3	153
54.5 - 61.5	58	2	116
61.5 - 68.5	65	9	585
68.5 - 75.5	72	13	936
75.5 - 82.5	79	8	632
82.5 - 89.5	86	3	258
89.5 - 96.5	93	2	186
		40	2866

a. $\overline{X} = \frac{\Sigma f \cdot X_m}{n} = \frac{2866}{40} = 71.65$

b. modal class: 68.5 - 75.5

15.

Class Limits	Boundaries	X_m	f	$f \cdot X_m$
14 - 20	13.5 - 20.5	17	10	170
21 - 27	20.5 - 27.5	24	11	264
28 - 34	27.5 - 34.5	31	6	186
35 - 41	34.5 - 41.5	38	8	304
42 - 48	41.5 - 48.5	45	4	180
49 - 55	48.5 - 55.5	52	1	52
			40	1156

15. continued

a. $\overline{X} = \dfrac{\sum f \cdot X_m}{n} = \dfrac{1156}{40} = 28.9$

b. modal class: 21 – 27

17.

Percentage	Boundaries	X_m	f	$f \cdot X_m$
15.2 - 19.6	15.15 - 19.65	17.4	3	52.2
19.7 - 24.1	19.65 - 24.15	21.9	15	328.5
24.2 - 28.6	24.15 - 28.65	26.4	19	501.6
28.7 - 33.1	28.65 - 33.15	30.9	6	185.4
33.2 - 37.6	33.15 - 37.65	35.4	7	247.8
37.7 - 42.1	37.65 - 42.15	39.9	0	0
42.2 - 46.6	42.15 - 46.65	44.4	1	44.4
			51	1359.9

a. $\overline{X} = \dfrac{\sum f \cdot X_m}{n} = \dfrac{1359.9}{51} = 26.66$ or 26.7

b. modal class: 24.2 – 28.6

19.

Boundaries	X_m	f	$f \cdot X_m$
0.5 - 19.5	10	12	120
19.5 - 38.5	29	7	203
38.5 - 57.5	48	5	240
57.5 - 76.5	67	3	201
76.5 - 95.5	86	3	258
		30	1022

a. $\overline{X} = \dfrac{\sum f \cdot X_m}{n} = \dfrac{1022}{30} = 34.1$

b. modal class: 0.5 – 19.5

21.

Children	f	$f \cdot X_m$
0	6	0
1	6	6
2	10	20
3	6	18
4	6	24
5	4	20
6	4	24
7	2	14
8	2	16
9	0	0
10	1	10
	47	152

a. $\overline{X} = \dfrac{\sum f \cdot X_m}{n} = \dfrac{152}{47} = 3.23$

b. mode: 2

23.

$\overline{X} = \dfrac{\sum w \cdot X}{\sum w} = \dfrac{8(10,000) + 10(12,000) + 12(8,000)}{8 + 10 + 12}$

$= \dfrac{296,000}{8 + 10 + 12}$

$= \dfrac{296,000}{30}$

$= \$9866.67$

25.

$\overline{X} = \dfrac{\sum w \cdot X}{\sum w} = \dfrac{40(1000) + 30(3000) + 50(800)}{1000 + 3000 + 800} = 35.4\%$

27.

$\overline{X} = \dfrac{\sum w \cdot X}{\sum w} = \dfrac{20(83) + 30(72) + 50(90)}{100} = 83.2$

29.

a. Mode   d. Mode

b. Median e. Mean

c. Median f. Median

Chapter 3 - Data Description

31.
Roman letters, $\overline{X}$
Greek letters, μ

33.
$5 \cdot 64 = 320$

35.
The mean of the original data is 30.
The means will be:
a. 40
b. 20
c. 300
d. 3
e. The results will be the same as adding, subtracting, multiplying, and dividing the mean by 10.

37.
a. $\sqrt[3]{(1.35)(1.24)(1.18)} = 1.2547 \approx 1.255$

Average growth rate: $1.255 - 1 = 0.255$ or 25.5%

b. $\sqrt[4]{(1.08)(1.06)(1.04)(1.05)} = 1.057397$

Average growth rate: $1.057 - 1 = 0.057$ or 5.7%

c. $\sqrt[5]{(1.10)(1.08)(1.12)(1.09)(1.03)} = 1.084$

Average growth rate: $1.084 - 1 = 0.084$ or 8.4%

d. $\sqrt[3]{(1.01)(1.03)(1.055)} = \sqrt[3]{1.0975165} = 1.032$

Average growth rate: $1.032 - 1 = 0.032$ or 3.2%

39. $MD = \frac{\frac{50}{2} - 0}{26}(3.7) + 0.75 = 4.31$

EXERCISE SET 3-2

1. The square root of the variance is equal to the standard deviation.

3. σ^2, σ

5. When the sample size is less than 30, the formula for the variance of the sample will underestimate the population variance.

7.
$R = 110.8 - 20.1 = 90.7$

$s^2 = \frac{n\sum X^2 - (\sum X)^2}{n(n-1)} = \frac{10(28,948.44) - 457.4^2}{10(10-1)}$

$= \frac{80,269.64}{90} = 891.9$

$s = \sqrt{891.88} = 29.9$

9.
Silver:
$R = 35.42 - 7.34 = 27.9$

$s^2 = \frac{n\sum X^2 - (\sum X)^2}{n(n-1)} = \frac{9(3998.77) - 172.46^2}{9(9-1)} = \frac{6246.45}{72}$

$= 86.75$

$s = \sqrt{86.8} = 9.314$

Tin:
$R = 15.75 - 4.83 = 10.92$

$s^2 = \frac{n\sum X^2 - (\sum X)^2}{n(n-1)} = \frac{9(1079.58) - 93.51^2}{9(9-1)} = \frac{972.14}{72}$

$= 13.5$

$s = \sqrt{13.5} = 3.67$

The prices of silver are more variable.

11.
Triplets:
$R = 7110 - 5877 = 1233$

$s^2 = \frac{n\sum X^2 - (\sum X)^2}{n(n-1)} = \frac{10(427,765,643) - 65267^2}{10(10-1)}$

$= \frac{17,875,141}{90} = 198,612.7$

Chapter 3 - Data Description

11. continued

$s = \sqrt{198,612.7} = 445.7$

Quadruplets:
$R = 512 - 345 = 167$

$s^2 = \frac{n\sum X^2 - (\sum X)^2}{n(n-1)} = \frac{10(1,925,217) - 4347^2}{10(10-1)}$

$= \frac{355,761}{90} = 3952.9$

$s = \sqrt{3952.9} = 62.9$

Quintuplets:
$R = 91 - 46 = 45$

$s^2 = \frac{n\sum X^2 - (\sum X)^2}{n(n-1)} = \frac{10(56,535) - 741^2}{10(10-1)} = \frac{16,269}{90}$

$= 180.8$

$s = \sqrt{180.8} = 13.4$

The data for triplets are most variable.

13.
$R = 46 - 26 = 20$
Using the range rule of thumb, $s \approx \frac{20}{4} = 5$

15.
$R = 580 - 283 = 297$

$s^2 = \frac{n\sum X^2 - (\sum X)^2}{n(n-1)} = \frac{8(1,552,471) - 3457^2}{8(8-1)} = 8373.6$

$s = \sqrt{8373.6} = 91.5$

17.
$R = 156 - 26 = 130$

$s^2 = \frac{n\sum X^2 - (\sum X)^2}{n(n-1)} = \frac{25(271,995) - 2471^2}{25(25-1)} = 1156.7$

$s = \sqrt{1156.7} = 34.0$

19.

X_m	f	$f \cdot X_m$	$f \cdot X_m^2$
10	2	20	200
13	20	260	3380
16	18	288	4608
19	7	133	2527
22	2	44	968
25	1	25	625
	50	770	12,308

$s^2 = \frac{n\sum f \cdot X_m^2 - (\sum f \cdot X_m)^2}{n(n-1)} = \frac{50(12,308) - 770^2}{50(50-1)} = 9.2$

$s = \sqrt{9.18} = 3.0$

21.

X_m	f	$f \cdot X_m$	$f \cdot X_m^2$
65	13	845	54,925
128	2	256	32,768
191	0	0	0
254	5	1270	322,580
317	1	317	100,489
380	1	380	144,400
443	0	0	0
506	1	506	256,036
569	2	1138	647,522
	25	4712	1,558,720

$s^2 = \frac{n\sum f \cdot X_m^2 - (\sum f \cdot X_m)^2}{n(n-1)} = \frac{25(1,558,720) - 4712^2}{25(25-1)}$

$= 27,941.76$

$s = \sqrt{27941.76} = 167.16$ or 167.2

Chapter 3 - Data Description

23.

X_m	f	$f \cdot X_m$	$f \cdot X_m^2$
5	5	25	125
14	7	98	1372
23	10	230	5290
32	3	96	3072
41	3	123	5043
50	2	100	5000
	30	672	19,902

$s^2 = \frac{n \sum f \cdot X_m^2 - (\sum f \cdot X_m)^2}{n(n-1)} = \frac{30(19,902) - 672^2}{30(30-1)} = 167.2$

$s = \sqrt{167.2} = 12.9$

25.

X_m	f	$f \cdot X_m$	$f \cdot X_m^2$
119	8	952	113,288
252	11	2772	698,544
385	2	770	296,450
518	1	518	268,324
651	4	2604	1,695,204
784	2	1568	1,229,312
	28	9184	4,301,122

$s^2 = \frac{n \sum f \cdot X_m^2 - (\sum f \cdot X_m)^2}{n(n-1)} = \frac{28(4,301,122) - 9184^2}{28(28-1)}$

$= 47,732.22$

$s = \sqrt{47,732.22} = 218.5$

27.

C. Var $= \frac{s}{X} = \frac{2.3}{11} = 0.209 = 20.9\%$

C. Var $= \frac{s}{X} = \frac{1.8}{8} = 0.225 = 22.5\%$

The factory workers' data are more variable.

29.

C. Var $= \frac{s}{X} = \frac{10.5}{80.2} = 0.131 = 13.1\%$

C. Var $= \frac{s}{X} = \frac{18.3}{120.6} = 0.152 = 15.2\%$

The waiting time for people who are discharged is more variable.

31.

a. $1 - \frac{1}{2^2} = \frac{3}{4}$ or 75%

b. $1 - \frac{1}{1.5^2} = 0.56$ or 56%

33.

$\frac{120}{160} = 0.75 = 75\%$ so $k = 2$

$72 + 2s = 77$

$s = 2.5$

$72 + 2.5k = 82$

$k = 4$

$1 - \frac{1}{4^2} = 0.9375$ or at least 93.75%.

35.

$1 - \frac{1}{k^2} = 0.8889 \qquad k = 3$

$\overline{X} = 3$ hours or 180 minutes and $s = 32$ minutes

$180 - 3(32) = 84$ minutes; $180 + 3(32) = 276$

At least 88.89% of the data values will fall between 84 and 276 minutes.

37.

$1 - \frac{1}{k^2} = 0.75 \qquad k = 2$

$\overline{X} = \$258,100$ and $s = \$48,500$

$\$258,100 - 2(\$48,500) = \$161,100$ and $\$258,100 + 2(\$48,500) = \$355,100$. At least 75% of the homes will fall between $161,100 and $355,100.

39.

$\overline{X} = 504$ and $s = 55.7$

$504 + 55.7k = 896.57$ so $k = 7.05$

$1 - \frac{1}{k^2} = 1 - \frac{1}{7.05^2} = 0.98$ or at least 98%

Chapter 3 - Data Description

41.
By the Empirical Rule, 68% of scores are within 1 standard deviation of the mean.
Thus, $538 - 1(48) = 490$ and $538 + 1(48) = 586$.
Therefore, 68% of the scores would fall between 490 and 586.

To find the percentage of scores above 634, first find k:
$538 + k(48) = 634$
$48k = 96$
$k = 2$
By the Empirical Rule, 95% of the data are within $k = 2$ standard deviations of the mean. This means that $100\% - 95\% = 5\%$ of the scores would be above and below 2 standard deviations of the mean. Thus, $\frac{1}{2}$ of 5% or 2.5% of the data are above 634.

43.
The average price of an instrument at a small music store is $325. The standard deviation of the price is $52. The Owner decides to raise the price of all the instruments by $20.
The new mean of prices is
$\overline{X} = \$325 + \$20 = \$345$, and the new standard deviation of prices is $s = \$52$.

45.
The mean price of the fish in a pet shop is $2.17, and the standard deviation of the price is $0.55.
The Owner decides to triple the prices.
The new mean of prices is
$\overline{X} = \$2.17 \times 3 = \$6.51$
and the new standard deviation of prices is
$s = \$0.55 \times 3 = \$1.65$.

47.
$n = 30 \quad \overline{X} = 215 \quad s = 20.8$
At least 75% of the data values will fall between $\overline{X} \pm 2s$.
$\overline{X} - 2(20.8) = 215 - 41.6 = 173.4$ and
$\overline{X} + 2(20.8) = 215 + 41.6 = 256.6$
In this case all 30 values fall within this range.

49.
For $k = 1.5$, $1 - \frac{1}{1.5^2} = 1 - 0.44$
$= 0.56$ or 56%
For $k = 2$, $1 - \frac{1}{2^2} = 1 - 0.25$
$= 0.75$ or 75%
For $k = 2.5$, $1 - \frac{1}{2.5^2} = 1 - 0.16$
$= 0.84$ or 84%
For $k = 3$, $1 - \frac{1}{3^2} = 1 - 0.1111$
$= .8889$ or 89%
For $k = 3.5$, $1 - \frac{1}{3.5^2} = 1 - 0.08$
$= 0.92$ or 92%

51.
$\overline{X} = 13.3$
Mean Dev
$= \frac{|5-13.3|+|9-13.3|+|10-13.3|+|11-13.3|+|11-13.3|}{10}$

$+ \frac{|12-13.3|+|15-13.3|+|18-13.3|+|20-13.3|+|22-13.3|}{10}$
$= 4.36$

53.
For $n = 25$, $\overline{X} = 50$, and $s = 3$:

$s\sqrt{n-1} = 3\sqrt{25-1} = 14.7$
$\overline{X} + s\sqrt{n-1} = 50 + 14.7 = 64.7$

67 may be an incorrect data value, since it is beyond the range using the formula $s\sqrt{n-1}$.

Chapter 3 - Data Description

EXERCISE SET 3-3

1. A z score tells how many standard deviations the data value is above or below the mean.

3. A percentile is a relative measure while a percent is an absolute measure of the part to the total.

5. $Q_1 = P_{25}$, $Q_2 = P_{50}$, $Q_3 = P_{75}$

7. $D_1 = P_{10}$, $D_2 = P_{20}$, $D_3 = P_{30}$, etc

9.
For Canada:
$z = \frac{X - \overline{X}}{s} = \frac{26 - 29.4}{8.6} = -0.40$
For Italy:
$z = \frac{X - \overline{X}}{s} = \frac{42 - 29.4}{8.6} = 1.47$
For US:
$z = \frac{X - \overline{X}}{s} = \frac{13 - 29.4}{8.6} = -1.91$

11.
a. $z = \frac{X - \overline{X}}{s} = \frac{27 - 24.6}{3.2} = 0.75$
b. $z = \frac{22 - 24.6}{3.2} = -0.8125$
c. $z = \frac{31 - 24.6}{3.2} = 2$
d. $z = \frac{18 - 24.6}{3.2} = -2.0625$
e. $z = \frac{26 - 24.6}{3.2} = 0.4375$

13.
For the geography test: $z = \frac{83 - 72}{6} = 1.83$

For the accounting test: $z = \frac{61 - 55}{3.5} = 1.71$

The geography test score is relatively higher than the accounting test score.

15.
a. $z = \frac{16{,}000 - 14{,}090}{3500} = 0.55$
b. $z = \frac{10{,}000 - 14{,}090}{3500} = -1.17$
c. To find the number of miles, use
$X = zs + \overline{X}$
$X = 1.6(3500) + 14{,}090 = 19{,}690$ miles
$X = -0.5(3500) + 14{,}090 = 12{,}340$ miles
$X = 0(3500) + 14{,}090 = 14{,}090$ miles

17.
a. For the 40th percentile:
$c = \frac{(27)(40)}{100} = 10.8$ or 11th value,
which is the data value of 21.

b. For the 75th percentile:
$c = \frac{(27)(75)}{100} = 20.25$ or 21st value,
which is the data value of 43.

c. For the 90th percentile:
$c = \frac{(27)(90)}{100} = 24.3$ or 25th value,
which is the data value of 97.

d. For the 30th percentile:
$c = \frac{(27)(30)}{100} = 8.1$ or 9th value,
which is the data value of 19.

a. For 27:
$P = \frac{15 + 0.5}{27}$
$= 0.574$ or the 57th percentile.

b. For 40:
$P = \frac{19 + 0.5}{27}$
$= 0.722$ or the 72nd percentile.

c. For 58:
$P = \frac{21 + 0.5}{27}$
$= 0.796$ or the 80th percentile.

17. continued

d. For 67:

$P = \dfrac{23 + 0.5}{27}$

$= 0.870$ or the 87th percentile.

19.

a. 6th b. 24th
c. 68th d. 76th
e. 94th f. 234
g. 251 h. 263
i. 274 j. 284

21.

Percentile $= \dfrac{\text{number of values below} + 0.5}{\text{total number of values}} \cdot 100\%$

Data: 228, 489, 524, 597, 623, 659, 736, 777, 804

For 228, $\dfrac{0+.5}{9} \cdot 100\% = 6^{th}$ percentile
For 489, $\dfrac{1+.5}{9} \cdot 100\% = 17^{th}$ percentile

For 524, $\dfrac{2+.5}{9} \cdot 100\% = 28^{th}$ percentile
For 597, $\dfrac{3+.5}{9} \cdot 100\% = 39^{th}$ percentile

For 623, $\dfrac{4+.5}{9} \cdot 100\% = 50^{th}$ percentile
For 659, $\dfrac{5+.5}{9} \cdot 100\% = 61^{st}$ percentile

For 736, $\dfrac{6+.5}{9} \cdot 100\% = 72^{nd}$ percentile
For 777, $\dfrac{7+.5}{9} \cdot 100\% = 83^{rd}$ percentile

For 804, $\dfrac{8+.5}{9} \cdot 100\% = 94^{th}$ percentile

$c = \dfrac{9(40)}{100} = 3.6$ or 4th data value, which is 597

23.

Percentile $= \dfrac{\text{number of values below} + 0.5}{\text{total number of values}} \cdot 100\%$

Data: 1.1, 1.7, 1.9, 2.1, 2.2, 2.5, 3.3, 6.2, 6.8, 20.3

23. continued

For 1.1, $\dfrac{0+.5}{10} \cdot 100\% = 5^{th}$ percentile
For 1.7, $\dfrac{1+.5}{10} \cdot 100\% = 15^{th}$ percentile

For 1.9, $\dfrac{2+.5}{10} \cdot 100\% = 25^{th}$ percentile
For 2.1, $\dfrac{3+.5}{10} \cdot 100\% = 35^{th}$ percentile

For 2.2, $\dfrac{4+.5}{10} \cdot 100\% = 45^{th}$ percentile
For 2.5, $\dfrac{5+.5}{10} \cdot 100\% = 55^{th}$ percentile

For 3.3, $\dfrac{6+.5}{10} \cdot 100\% = 65^{th}$ percentile
For 6.2, $\dfrac{7+.5}{10} \cdot 100\% = 75^{th}$ percentile

For 6.8, $\dfrac{8+.5}{10} \cdot 100\% = 85^{th}$ percentile
For 20.3, $\dfrac{9+.5}{10} \cdot 100\% = 95^{th}$ percentile

$c = \dfrac{10(40)}{100} = 4$

average the 4th and 5th values:

$P_{40} = \dfrac{2.1 + 2.2}{2} = 2.15$

25.

To find Q_1, find P_{25}:

$c = \dfrac{(10)(25)}{100} = 2.5$, round up to 3.

Q_1 is at the 3rd value, which is 11.

To find Q_3, find P_{75}:

$c = \dfrac{(10)(75)}{100} = 7.5$, round up to 8.

Q_3 is at the 8th value, which is 32.
IQR $= Q_3 - Q_1 = 32 - 11 = 21$.

27.

To find Q_1, find P_{25}:

$c = \dfrac{(11)(25)}{100} = 2.75$, round up to 3.

Q_1 is at the 3rd value, which is 19.7.

To find Q_3, find P_{75}:

$c = \dfrac{(11)(75)}{100} = 8.25$, round up to 9.

27. continued

Q_3 is at the 9th value, which is 78.8.
IQR = $Q_3 - Q_1$ = 78.8 − 19.7 = 59.1.

29.

a. 19 21 25 28 29 32 34 46
          ↑   ↑   ↑
          Q_1 MD Q_3

MD = $\frac{28+29}{2}$ = 28.5

For Q_1: $Q_1 = \frac{21+25}{2} = 23$
For Q_3: $Q_3 = \frac{32+34}{2} = 33$

$Q_3 - Q_1$ = 33 − 23 = 10 and 10(1.5) = 15.
23 − 15 = 8 and 33 + 15 = 48.

Since all the values fall within the range of 8 to 48, there are no outliers.

b. 65 82 89 90 93 94 97 100 101
    ↑   ↑   ↑
    Q_1 MD Q_3

MD = 93

For Q_1: $Q_1 = \frac{82+89}{2} = 85.5$
For Q_3: $Q_3 = \frac{97+100}{2} = 98.5$

$Q_3 - Q_1$: 98.5 − 85.5 = 13 and
13(1.5) = 19.5. 85.5 − 19.5 = 66 and
98.5 + 19.5 = 118.
Only the value 65 lies outside the range of 66 to 118 and is a suspected outlier.

c. 175 371 489 527 1007
      ↑   ↑   ↑
      Q_1 MD Q_3

MD = 489

For Q_1: $Q_1 = \frac{175+371}{2} = 273$
For Q_3: $Q_3 = \frac{527+1007}{2} = 767$

29. continued

$Q_3 - Q_1$: 767 − 273 = 494 and
494(1.5) = 741.
273 − 741 = −468
and 767 + 741 = 1508.

Since all the values fall within the range of −468 to 1508, there are no outliers.

31.

a. 5, 12, 16, 25, 32, 38
$Q_1 = 12$, $Q_2 = 20.5$, $Q_3 = 32$
Midquartile = $\frac{12+32}{2} = 22$
Interquartile range: 32 − 12 = 20

b. 53, 62, 78, 94, 96, 99, 103
$Q_1 = 62$, $Q_2 = 94$, $Q_3 = 99$
Midquartile = $\frac{62+99}{2} = 80.5$
Interquartile range: 99 − 62 = 37

33.

Tom's score is 158. Harry's score can be calculated based on his z score:
$X = 2(18) + 125 = 161$.

Since the data are normally distributed, 95% fall within 2 standard deviations of the mean (using the Empirical Rule).
Thus, $125 \pm 2(18)$ gives a range of 89 to 161, and a score of 161 is the 95th percentile. Since Dick scored in the 98th percentile, his raw score must be higher than 161.

Therefore, Tom's score is the lowest followed by Harry, with Dick's score being the highest.

Chapter 3 - Data Description

EXERCISE SET 3-4

1. Data arranged in order:
6, 8, 12, 19, 27, 32, 54
Minimum: 6
Q_1: 8
Median: 19
Q_3: 32
Maximum: 54
Interquartile Range: $32 - 8 = 24$

3. Data arranged in order:
188, 192, 316, 362, 437, 589
Minimum: 188
Q_1: 192
Median: $\frac{316 + 362}{2} = 339$
Q_3: 437
Maximum: 589
Interquartile Range: $437 - 192 = 245$

5. Data arranged in order:
14.6, 15.5, 16.3, 18.2, 19.8
Minimum: 14.6
Q_1: $\frac{14.6 + 15.5}{2} = 15.05$
Median: 16.3
Q_3: $\frac{18.2 + 19.8}{2} = 19.0$
Maximum: 19.8
Interquartile Range: $19.0 - 15.05 = 3.95$

7. Minimum: 3
Q_1: 5
Median: 8
Q_3: 9
Maximum: 11
Interquartile Range: $9 - 5 = 4$

9. Minimum: 55
Q_1: 65
Median: 70
Q_3: 90
Maximum: 95
Interquartile Range: $90 - 65 = 25$

11.
MD = 82 $Q_1 = 79$ $Q_3 = 89$

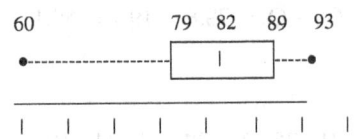

The distribution is right-skewed.

13.
MD = $\frac{14 + 18}{2} = 16$ $Q_1 = 5.8$
$Q_3 = 27.8$

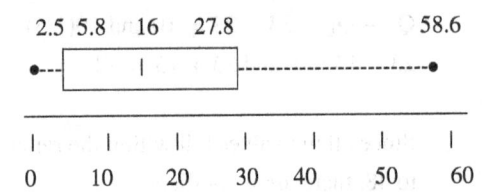

The box plot of the data is somewhat positively skewed.

15.
For Baltic Sea:
MD = 1154 $Q_1 = \frac{228 + 610}{2} = 419$
$Q_3 = \frac{1159 + 2772}{2} = 1965.5$

For Aleutian Islands:
MD = 686 $Q_1 = \frac{275 + 350}{2} = 312.5$
$Q_3 = \frac{1051 + 1571}{2} = 1311$

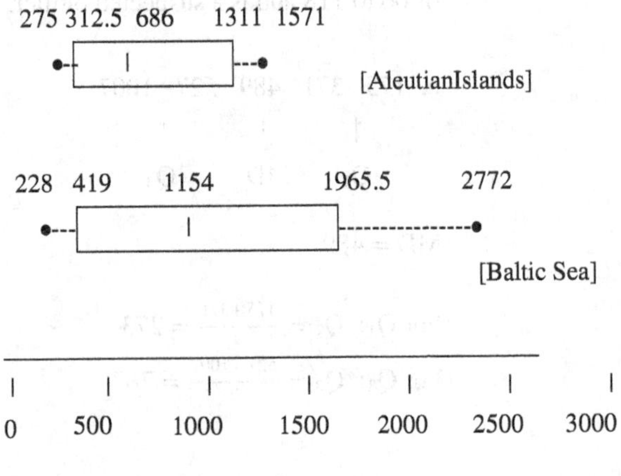

Chapter 3 - Data Description

15. continued

The areas of the islands in the Baltic Sea are more variable than the ones in the Aleutian Islands. Also, they are in general larger in area.

17.

$MD = \frac{606 + 645}{2} = 625.5$ $Q_1 = 585$

$Q_3 = 717$

Lowest value = 564   Highest value = 770

IQR = 132

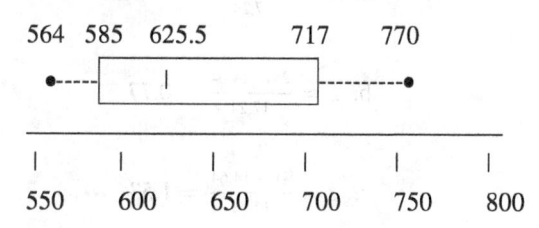

564 585 625.5   717   770

| | | | | |
550 600 650 700 750 800

19. Data arranged in order: 39, 39, 42, 43, 43, 53, 54, 66, 91, 97

Minimum: 39

Q_1: 42

Median: $\frac{43 + 53}{2} = 48$

Q_3: 66

Maximum: 97

Interquartile Range: $66 - 42 = 24$

$1.5(24) = 36$ for mild outliers;
$3(24) = 72$ for extreme outliers
There are no outliers.

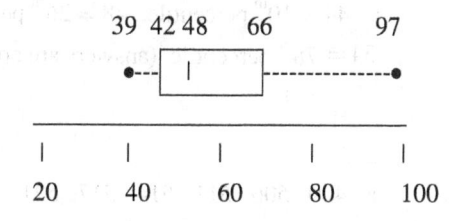

39 42 48 66   97

| | | | |
20   40   60   80   100

REVIEW EXERCISES - CHAPTER 3

1.

$\overline{X} = \frac{\sum X}{n} = \frac{548}{15} = 36.5$

1. continued

Data arranged in order: 0, 0, 3, 3, 4, 4, 7, 11, 14, 24, 30, 51, 92, 148, 157

MD = 11

Mode = 0, 3, and 4

$MR = \frac{0 + 157}{2} = 78.5$

3.

Class	X_m	f	$f \cdot X_m$
105 - 109	107	2	214
110 - 114	112	5	560
115 - 119	117	6	702
120 - 124	122	8	976
125 - 129	127	8	1016
130 - 134	132	1	132
		30	3600

$\overline{X} = \frac{\sum f \cdot X_m}{n} = \frac{3600}{30} = 120$

Modal Classes = 120 - 124 or 119.5 - 124.5 and 125 - 129 or 124.5 - 129.5

5.

$\overline{X} = \frac{\sum w \cdot X}{\sum w} = \frac{1.6(1.4) + 0.8(0.8) + 0.4(0.3) + 1.8(1.6)}{1.4 + 0.8 + 0.3 + 1.6}$

$= 1.43$ viewers per household

7.

Range = $212 - 37 = 175$

$s^2 = \frac{n \sum f \cdot X_m^2 - (\sum f \cdot X_m)^2}{n(n-1)} = \frac{12(110,077) - 989^2}{12(12-1)}$

$= 2596.99$

$s = \sqrt{2596.99} = 51.0$

Chapter 3 - Data Description

9.

Class Boundaries	X_m	f	$f \cdot X_m$	$f \cdot X_m^2$	cf
12.5 - 27.5	20	6	120	2400	6
27.5 - 42.5	35	3	105	3675	9
42.5 - 57.5	50	5	250	12,500	14
57.5 - 72.5	65	8	520	33,800	22
72.5 - 87.5	80	6	480	38,400	28
87.5 - 102.5	95	2	190	18,050	30
		30	1665	108,825	

a. $\overline{X} = \dfrac{\sum f \cdot X_m}{n} = \dfrac{1665}{30} = 55.5$

b. Modal class = 57.5 – 72.5

c. $s^2 = \dfrac{n\sum f \cdot X_m^2 - (\sum f \cdot X_m)^2}{n(n-1)} = \dfrac{30(108,825) - 1665^2}{30(30-1)}$
$= 566.1$

d. $s = \sqrt{566.1} = 23.8$

11. $s \approx \dfrac{24}{4} = 6$

13.
Textbooks: C. Var $= \dfrac{5}{16} = 0.3125$ or 31.25%
Ages: C. Var $= \dfrac{8}{43} = 0.186$ or 18.6%
The number of books is more variable.

15.
$\overline{X} = 0.32 \quad s = 0.03 \quad k = 2$
$0.32 - 2(0.03) = 0.26$
and $0.32 + 2(0.03) = 0.38$
At least 75% of the values
will fall between $0.26 and $0.38.

17.
$\overline{X} = 54 \quad s = 4 \quad 60 - 54 = 6$
$k = \dfrac{6}{4} = 1.5$
$1 - \dfrac{1}{1.5^2} = 1 - 0.44$
$= 0.56$ or 56%

19. By the Empirical Rule, 68% of the scores will be within 1 standard deviation of the mean.
$21 + 1(4) = 25$
$21 - 1(4) = 17$
Then, 68% of the haircut cost is between $17 and $25.

21.
$\overline{X} = 14.64$
$s = 17.24$

a. $z = \dfrac{10 - 14.64}{17.24} = -0.27$

b. $z = \dfrac{28 - 14.64}{17.24} = 0.77$

c. $z = \dfrac{41 - 14.64}{17.24} = 1.53$

23.
a.

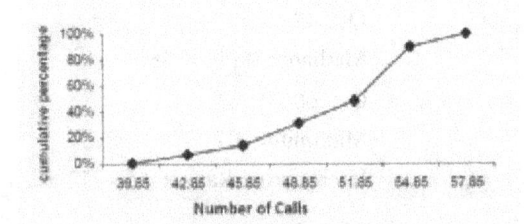

b. $P_{35} = 50; \; P_{65} = 53; \; P_{85} = 55$
(answers are approximate)

c. $44 = 10^{th}$ percentile; $48 = 26^{th}$ percentile; $54 = 78^{th}$ percentile (answers are approximate)

25.
a. 400 506 511 514 517 521
         ↑   ↑
         Q_1 Q_3

For Q_1: $c = \dfrac{np}{100} = \dfrac{6(25)}{100} = 1.5$ round up to 2
$Q_1 = 506$

Chapter 3 - Data Description

25. continued

For Q_3: $c = \frac{np}{100} = \frac{6(75)}{100} = 4.5$ round up to 5

$Q_3 = 517$

$Q_3 - Q_1 = 517 - 506 = 11$; $11(1.5) = 16.5$;
$506 - 16.5 = 489.5$ and $517 + 16.5 = 533.5$
Therefore, only the value 400 lies outside the range of 489.5 to 533.5 and is a suspected outlier.

b. 3 6 7 8 9 10 12 14 16 20
          ↑   ↑
         Q_1 Q_3

For Q_1: $c = \frac{np}{100} = \frac{10(25)}{100} = 2.5$ round up to 3

$Q_1 = 7$

For Q_3: $c = \frac{np}{100} = \frac{10(75)}{100} = 7.5$ round up to 8

$Q_3 = 14$

$Q_3 - Q_1 = 14 - 7 = 7$; $7(1.5) = 10.5$;
$7 - 10.5 = -3.5$ and $14 + 10.5 = 24.5$
Since all values fall within the range of -3.5 to 24.5, there are no outliers.

27.

For years 1851–1860:
MD = 6 $Q_1 = 5$ $Q_3 = 7$

For 1941–1950:
MD = 10 $Q_1 = 9$ $Q_3 = 11$

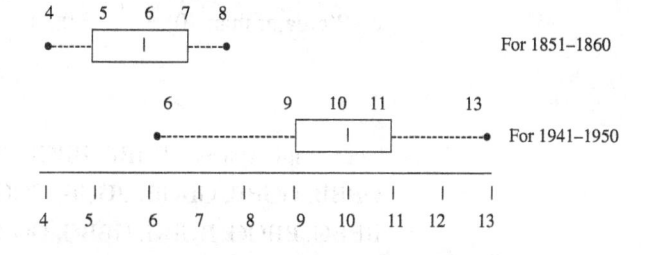

```
  4   5   6   7   8
  •---[---|---]---•                    For 1851–1860

          6        9  10  11    13
          •--------[--|--]--------•    For 1941–1950

  |---|---|---|---|---|---|---|---|
  4   5   6   7   8   9  10  11  12  13
```

The data for years 1941–1950 have a higher median and are more variable.

CHAPTER 3 QUIZ

1. True
3. False
5. False
7. False
9. False
11. c
13. b
15. b
17. Parameters, statistics
19. σ
21. Positively
23. a. 15.3   b. 15.5
    c. 15, 16, 17   d. 15
    e. 6   f. 3.57
    g. 1.9
25. 4.46 or 4.5
27. 88.89%
29. $s \approx \frac{18}{4} = 4.5$
31.
  a.

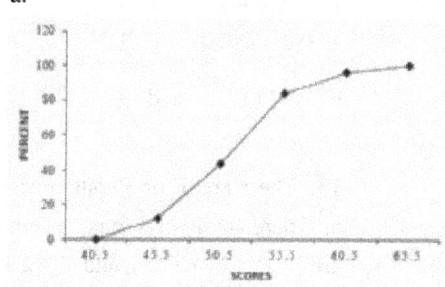

  b. 47; 55; 64

  c. 56th percentile; 6th percentile; 99th percentile

Chapter 4 - Probability and Counting Rules

Note: Answers may vary due to rounding, TI-83's or computer programs.

EXERCISE SET 4-1

1. A probability experiment is a chance process which leads to well-defined outcomes.

3. An outcome is the result of a single trial of a probability experiment, whereas an event can consist of one or more outcomes.

5. The range of values is $0 \leq P(E) \leq 1$.

7. 0

9.
$1 - 0.20 = 0.80$
Since the probability that it won't rain is 80%, you could leave your umbrella at home and be fairly safe.

11.
a. Empirical   c. Empirical
b. Classical   d. Classical

13.
a. 0   c. 1
b. $\frac{1}{2}$ d. $\frac{1}{2}$

15. There are 6^2 or 36 outcomes.
a. There are 4 ways to get a sum of 5. They are (4,1), (3,2), (2,3), and (1,4). The probability then is $\frac{4}{36} = \frac{1}{9}$.

b. There are 4 ways to get a sum of 9 and 3 ways to get a sum of 10. They are (6,4), (5,5), (4,6), (6,3), (5,4), (4,5), and (3,6). The probability then is $\frac{7}{36}$.

c. There are 6 ways to get doubles. They are (1,1), (2,2), (3,3), (4,4), (5,5), and (6,6). The probability then is $\frac{6}{36} = \frac{1}{6}$.

17.
a. $\frac{1}{13}$ d. $\frac{2}{13}$
b. $\frac{1}{4}$ e. $\frac{6}{13}$
c. $\frac{1}{52}$

19. There are 20 possible outcomes.

a. P(winning $10) = P(rolling a 1)
P(rolling a 1) $= \frac{2}{20} = \frac{1}{10} = 0.1$

b. P(winning $5 or $10) = P(rolling either a 1 or 2)
P(1 or 1) $= \frac{4}{20} = \frac{1}{5} = 0.2$

c. P(winning a coupon) = P(rolling either a 3 or 4)
P(3 or 4) $= \frac{16}{20} = \frac{4}{5} = 0.8$

21.
a. P(type B) $= 0.12$ or 12%

b. P(type AB or O) $= 0.05 + 0.43 = 0.48$ or 48%

c. P(not type O) $= 1 - P(\text{type O})$
$= 1 - 0.43$
$= 0.57$ or 57%

23.
a. P(odd prime number) $= \frac{24}{25} = 0.96$

b. P(sum of the digits is odd) $= \frac{12}{25} = 0.48$

c. P(greater than 70) $= \frac{6}{25} = 0.24$

25.
The sample space is BBBB, BBGB, BGBB, GBBB, GGBB, GBGB, BGGB, GGGB, BBBG, BBGG, BGBG, GBBG, GGBG, GBGG, BGGG, and GGGG.

a. All girls is the outcome GGGG; hence P(all girls) $= \frac{1}{16}$.

Chapter 4 - Probability and Counting Rules

25. continued

b. Exactly two girls and two boys would be GGBB, GBGB, BGGB, BBGG, BGBG, GBBG; hence, P(exactly two girls and two boys) = $\frac{6}{16} = \frac{3}{8}$.

c. At least one child who is a girl would be all outcomes, apart from BBBB. The probability then is $\frac{15}{16}$.

d. At least one child of each gender means at least one boy or at least one girl. The outcomes are BBGB, BGBB, GBBB, GGBB, GBGB, BGGB, GGGB, BBBG, BBGG, BGBG, GBBG, GGBG, GBGG, BGGG. Hence the probability is $\frac{14}{16} = \frac{7}{8}$.

27.

The outcomes for 7 or 11 are (1,6), (2,5), (3,4), (4,3), (5,2), (6,1), (5,6), and (6,5); hence, P(7 or 12) = $\frac{8}{36}$.

The outcomes for 2, 3, or 12 are (1,1), (1,2), (2,1), and (6,6); hence, P(2, 3, or 12) = $\frac{1+2+1}{36} = \frac{4}{36}$.

P(game will last only one roll) = $\frac{8}{36} + \frac{4}{36}$
$= \frac{12}{36} = \frac{1}{3}$.

29.

a. P(debt is less than $5001) = 27%.

b. P(debt is more than $20,000) = P($20,001 to $50,000) + P($50,000+) = 19% + 14% = 33%

c. P(debt is between $1 and $20,000) = P($1 to $5000) + P($5001 to $20,000) = 27% + 40% = 67%

d. P(debt is more than $50,000) = 14%

31.

P(motor vehicle theft) = $\frac{275}{2500} = 0.11$

P(not an assault) = 1 − P(assault)

P(not an assault) = $1 - \frac{200}{2500} = 0.92$

33.

P(either a truck or a motorcycle accident) = $\frac{5,200,000 + 178,000}{18,878,000} = 0.285$

P(not a truck accident) = 1 − P(truck accident) = $1 - \frac{5,200,000}{18,878,000} = 0.725$

35.

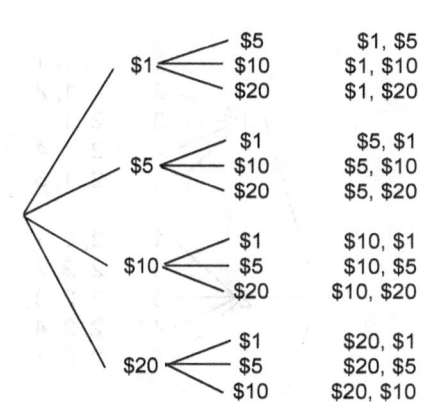

```
        $5       $1, $5
   $1 < $10      $1, $10
        $20      $1, $20

        $1       $5, $1
   $5 < $10      $5, $10
        $20      $5, $20

        $1       $10, $1
  $10 < $5       $10, $5
        $20      $10, $20

        $1       $20, $1
  $20 < $5       $20, $5
        $10      $20, $10
```

37.

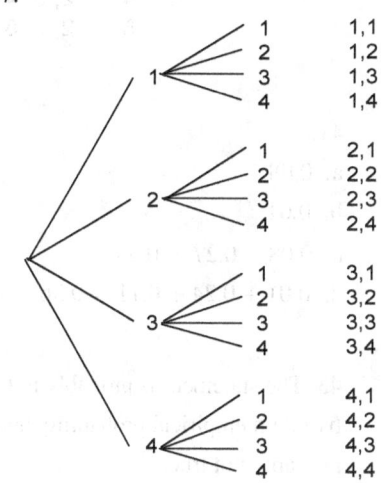

```
      1    1,1
   1< 2    1,2
      3    1,3
      4    1,4

      1    2,1
   2< 2    2,2
      3    2,3
      4    2,4

      1    3,1
   3< 2    3,2
      3    3,3
      4    3,4

      1    4,1
   4< 2    4,2
      3    4,3
      4    4,4
```

Chapter 4 - Probability and Counting Rules

39.

English Math Elective

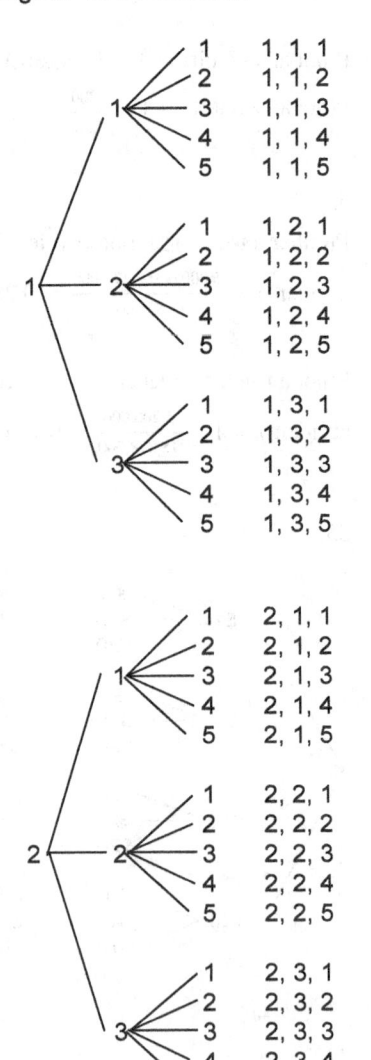

```
         1       1, 1, 1
         2       1, 1, 2
    1 ─  3       1, 1, 3
         4       1, 1, 4
         5       1, 1, 5

         1       1, 2, 1
         2       1, 2, 2
1 ─ 2 ─  3       1, 2, 3
         4       1, 2, 4
         5       1, 2, 5

         1       1, 3, 1
         2       1, 3, 2
    3 ─  3       1, 3, 3
         4       1, 3, 4
         5       1, 3, 5

         1       2, 1, 1
         2       2, 1, 2
    1 ─  3       2, 1, 3
         4       2, 1, 4
         5       2, 1, 5

         1       2, 2, 1
         2       2, 2, 2
2 ─ 2 ─  3       2, 2, 3
         4       2, 2, 4
         5       2, 2, 5

         1       2, 3, 1
         2       2, 3, 2
    3 ─  3       2, 3, 3
         4       2, 3, 4
         5       2, 3, 5
```

41.

a. 0.08

b. 0.01

c. $0.08 + 0.27 = 0.35$

d. $0.01 + 0.24 + 0.11 = 0.36$

43. The statement is probably not based on empirical probability and probably not true.

45. Actual outcomes will vary, however the probabilities of 0, 1, 2, or 3 heads should be approximately $\frac{1}{8}$, $\frac{3}{8}$, $\frac{3}{8}$, and $\frac{1}{8}$ respectively.

47.

a. 1:5, 5:1   e. 1:12, 12:1

b. 1:1, 1:1   f. 1:3, 3:1

c. 1:3, 3:1   g. 1:1, 1:1

d. 1:1, 1:1

EXERCISE SET 4-2

1. Two events are mutually exclusive if they cannot occur at the same time. Examples will vary.

3.

a. Not mutually exclusive

   You can get the 6 of spades.

b. Mutually exclusive

c. Mutually exclusive

d. Not mutually exclusive

   Some sophomore students are male.

5.

a. $\frac{1,348,503}{1,907,172} = 0.707$

b. $\frac{46,024}{1,907,172} + \frac{1,098,371}{1,907,172} - \frac{21,683}{1,907,172} =$

$\frac{1,122,712}{1,907,172} = 0.589$

c. $\frac{21,683}{1,907,172} = 0.011$

d. $\frac{1,394,527}{1,907,172} = 0.731$

7.

a. P(pathologist) = $\frac{7}{38}$ or 0.184

b. P(orthopedist or MD) = $\frac{22}{38} + \frac{33}{38} - \frac{20}{38}$

P(orthopedist or MD) = $\frac{35}{38}$ or 0.921

9.

$\frac{8}{16} + \frac{2}{16} = \frac{10}{16} = \frac{5}{8}$

Chapter 4 - Probability and Counting Rules

11.

	Cheese Pizzas	Pizzas with one or more toppings	Total
Eaten at work	12	10	22
Not eaten at work	12	6	18
Total	24	16	40

a. P(a cheese pizza eaten at work) $= \frac{12}{40} = \frac{3}{10} = 0.30$

b. P(a pizza with either one or more toppings, and it was not eaten at work) $= \frac{3}{4} = 0.75$

c. P(a cheese pizza or a pizza eaten at work)
$= \frac{24}{40} + \frac{22}{40} - \frac{12}{40} = \frac{34}{40} = \frac{17}{20} = 0.85$

13.

	18 - 24	25 - 34	Total
Male	7922	2534	10,456
Female	5779	995	6,774
Total	13,701	3529	17,230

a. P(female aged 25 - 34) $= \frac{995}{17,230} = 0.058$

b. P(male or aged 18 - 24) =

$\frac{10,456}{17,230} + \frac{13,701}{17,230} - \frac{7922}{17,230} = \frac{16,235}{17,230} = 0.942$

c. P(under 25 years and not male) =

$\frac{5779}{17,230} = 0.335$

15.
Total $= 136,238$ multiple births

a. P(more than two babies) =
$\frac{7663}{136,328} = 0.056$

b. P(quads or quints) $= \frac{553}{136,328} = 0.004$

15. continued

c. The total number of babies who are triplets $= 21,330$
The total number of babies from multiple births $= 280,957$

P(baby is a triplet) $= \frac{21,330}{280,957} = 0.076$

17.

Age	High School	College	Neither	Total
Under 30	53	107	450	610
30 and over	27	32	367	426
Total	80	139	817	1036

a. P(The prisoner does not take classes) $= \frac{817}{1036} = 0.789$

b. P(under 20 and is taking either a high school class or a college class)
$= \frac{53}{1036} + \frac{107}{1036} = 0.154$

c. P(over 30 and is taking either a high school class or a college class)
$= \frac{27}{1036} + \frac{32}{1036} = 0.057$

19.
The total of the frequencies is 30.

a. $\frac{2}{30} = \frac{1}{15}$

b. $\frac{2+3+5}{30} = \frac{10}{30} = \frac{1}{3}$

c. $\frac{12+8+2+3}{30} = \frac{25}{30} = \frac{5}{6}$

d. $\frac{12+8+2+3}{30} = \frac{25}{30} = \frac{5}{6}$

e. $\frac{8+2}{30} = \frac{10}{30} = \frac{1}{3}$

21.
The total of the frequencies is 30.

a. $\frac{4}{30} = \frac{2}{15}$

21. continued

b. $\frac{11+9+5}{30} = \frac{25}{30} = \frac{5}{6}$

c. $\frac{9}{30} + \frac{5}{30} = \frac{14}{30} = \frac{7}{15}$

d. $\frac{11+9+5}{30} = \frac{25}{30} = \frac{5}{6}$

e. $\frac{4+1}{30} = \frac{5}{30} = \frac{1}{6}$

23.

a. There are 4 sevens, 4 eights, and 4 nines; hence, P(seven or eight or nine) $= \frac{12}{52} = \frac{3}{13}$

b. There are 13 spades, 4 kings, and 4 queens, but the king and queen of spades were counted twice.
Hence, P(spade or king or queen) = P(spade) + P(king) + P(queen) − P(king and queen of spades) $= \frac{13}{52} + \frac{4}{52} + \frac{4}{52} - \frac{2}{52} = \frac{19}{52}$

c. There are 13 clubs, and 12 face cards, but the face card of clubs was counted twice.
Hence, P(club or face) = P(club) + P(face) − P(face card of clubs) $= \frac{13}{52} + \frac{12}{52} - \frac{3}{52} = \frac{22}{52} = \frac{11}{26}$

d. There are 4 aces, 13 diamonds, and 13 hearts. There is one ace of diamonds and one ace of hearts.
Hence, P(ace or diamond or heart) = P(ace) + P(diamond) + P(heart) − P(ace of hearts and ace of diamonds) $= \frac{4}{52} + \frac{13}{52} + \frac{13}{52} - \frac{2}{52} = \frac{28}{52} = \frac{7}{13}$

23. continued

e. There are 4 nines, 4 tens, 13 spades, and 13 clubs. There is one nine of spades, one ten of spades, one nine of clubs and one ten of clubs. Hence, P(9 or 10 or spade or club) = P(9) + P(10) + P(spades) + P(club) − P(9 and 10 of clubs and spades)
$= \frac{4}{52} + \frac{4}{52} + \frac{13}{52} + \frac{13}{52} - \frac{4}{52} = \frac{30}{52} = \frac{15}{26}$

25.

P(apple juice or apple sauce) $= \frac{4.4}{11} + \frac{1}{11}$
$= 0.491$

27.

P(mushrooms or pepperoni) =
P(mushrooms) + P(pepperoni) −
P(mushrooms and pepperoni)

Let X = P(mushrooms and pepperoni)
Then $0.55 = 0.32 + 0.17 - X$
$X = 0.06$

29.

P(not a two-car garage) $= 1 - 0.70 = 0.30$

31.

$P(A \text{ or } B) = \frac{m}{2m+n} + \frac{n}{2m+n}$

$P(A \text{ or } B) = \frac{m+n}{2m+n}$

EXERCISE SET 4-3

1.
a. Independent   c. Dependent
b. Dependent   d. Dependent

3.
a. P(none play video or computer games) $= (0.31)^4 = 0.009$ or 0.9%

3. continued

P(all four play video or computer games) = $(0.69)^4 = 0.227$ or 22.7%

5.

P(making a sale) = 0.21

P(making 4 sales) = $(0.21)^4 = 0.0019$ or 0.002

The event is unlikely to occur since the probability is small.

7.

a. If 66% of law enforcement workers are sworn officers, and 88.4% of those workers are male, then $100\% - 88.4\% = 11.6\%$ of 66% are female sworn officers. Thus, P(female sworn officer)
= 0.116(66%) = 7.656% or 0.07656

b. If 66% of law enforcement workers are sworn officers, then $100\% - 66\% = 34\%$ are civilian workers, both male and female. Likewise, if 60.7% of civilian workers are female, then $100\% - 60.7\% = 39.3\%$ are male civilian workers. Thus, P(male civilian employee) = 0.393(34%) = 13.362% or 0.13362

c. The total of male employees, both sworn officers and civilian, is 71.706%. The total of civilian employees is 34%. Thus,
P(male or civilian) = P(male) + P(civilian) − P(both)
P(male or civilian) = 0.71706 + 0.34 − 0.13362
P(male or civilian) = 0.92344

9.

P(none are mothers) = $(0.25)^3 = 0.016$

11.

a. P(none of the three households had a smart TV) = $(1 - 0.45)^3 = 0.166375$

b. P(all three households had a smart TV) = $(0.45)^3 = 0.091125$

c. P(at least one of the three households had a smart TV)
= 1 − 0.166375
= 0.833625

13.

a. $\frac{4}{52} \cdot \frac{3}{51} \cdot \frac{2}{50} \cdot \frac{1}{49} = \frac{1}{270,725} = 0.00000369$

b. $\frac{26}{52} \cdot \frac{25}{51} \cdot \frac{24}{50} \cdot \frac{23}{49} = \frac{358800}{6497400} = \frac{46}{833} = 0.055$

c. $\frac{13}{52} \cdot \frac{12}{51} \cdot \frac{11}{50} \cdot \frac{10}{49} = \frac{17160}{6,497,400} = \frac{11}{4165} = 0.00264$

15.

a. P(both are nines) = $\frac{4}{52} \cdot \frac{3}{51} = \frac{1}{221}$

b. P(both are the same suit) = $\frac{4}{4} \cdot \frac{12}{51} = \frac{4}{17}$

c. P(both are spades) = $\frac{13}{52} \cdot \frac{12}{51} = \frac{1}{17}$

17.

P(both are dead) = $\frac{2}{12} \cdot \frac{1}{11} = \frac{1}{66} \approx 0.015$

Highly unlikely

19.

```
                                    Fed. Aid (0.448)
                        Aid (0.606)<
        Male (0.4237) <              No Fed. Aid (0.552)
                        No Aid (0.394)

                                    Fed. Aid (0.504)
                        Aid (0.652) <
        Female (0.5763)<             No Fed. Aid (0.496)
                        No Aid (0.348)
```

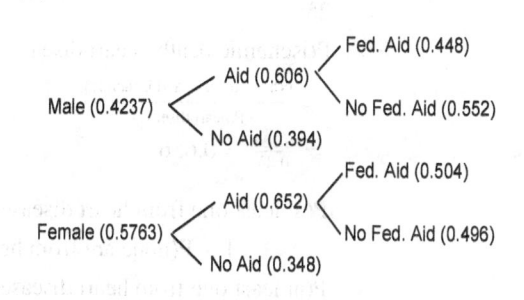

19. continued

a. P(male student without aid) $= 0.4237(0.394) = 0.0167$

b. P(male student | student has aid) $=$
$$\frac{P(\text{aid \& male})}{P(\text{aid})} = \frac{0.4237(0.606)}{0.4237(0.606)+0.5763(0.652)} = 0.406$$

c. P(female student or a student who receives federal aid) $=$
P(female) + P(federal aid) − P(female with federal aid) $=$
$0.5763 + (0.115 + 0.1894) − 0.1894 = 0.69$

21.

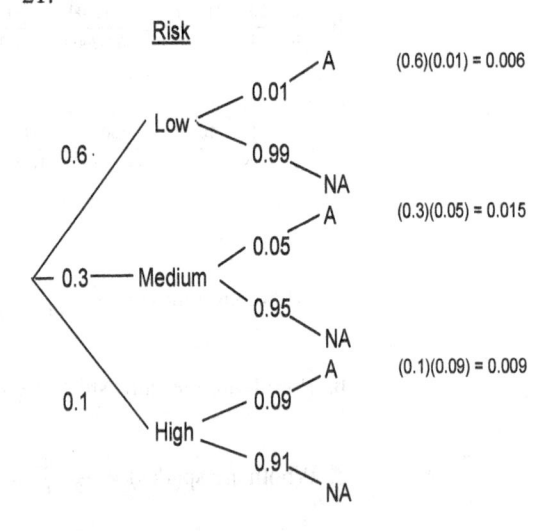

Risk

- 0.6 — Low
  - 0.01 → A: $(0.6)(0.01) = 0.006$
  - 0.99 → NA
- 0.3 — Medium
  - 0.05 → A: $(0.3)(0.05) = 0.015$
  - 0.95 → NA
- 0.1 — High
  - 0.09 → A: $(0.1)(0.09) = 0.009$
  - 0.91 → NA

P(accident) $= .006 + .015 + .009 = 0.03$

23.

P(female | adult) $= \frac{0.07}{0.99} = 0.0.071$

25.

P(ischemic death | heart disease)
$$= \frac{P(\text{heart disease and ischemic})}{P(\text{heart disease})}$$
$= \frac{0.164}{0.25} = 0.656$

P(at least one from heart disease) $=$
 $1 −$ P(none are from heart disease)
P(at least one from heart disease) $=$
 $1 − 0.75^2 = 0.4375$ or 0.438

27.

P(calculus | dean's list) $= \frac{0.042}{0.21} = 0.2$

29.

P(salad | pizza) $= \frac{0.65}{0.95} = 0.684$ or 68.4%

31.

a. $P(O^-) = 0.06$

b. $P(\text{type O} \mid Rh^+) = \frac{0.37}{0.85} = 0.435$

c. $P(A^+ \text{ or } AB^-) = 0.34 + 0.01 = 0.35$

d. $P(Rh^- \mid \text{type B}) = \frac{0.02}{0.12} = 0.167$

33.

a. P(tree | after 2000) $= \frac{\frac{77}{623}}{\frac{388}{623}} = 0.198$

b. P(camping and before 2001) $= \frac{117}{623} = 0.188$

c. P(camping | before 2001) $= \frac{\frac{117}{623}}{\frac{235}{623}} = 0.498$

35.

a. P(all 3 get enough exercise) $= (0.27)^3 = 0.0197$ or 0.020

b. P(at least one gets enough exercise) $= 1 − (0.73)^3 = 0.611$

37.

a. P(none have been married) $= (0.703)^5 = 0.172$

b. P(at least one has been married) $=$
$1 −$ P(none have been married)
$= 1 − 0.1717$
$= 0.828$

39.

P(at least one not on time)
$= 1 −$ P(none not on time)
$= 1 −$ P(all 5 on time)
$= 1 − (0.855)^5 = 0.5431$

Chapter 4 - Probability and Counting Rules

41.

If P(read to) = 0.58, then
P(not being read to) = 1 − 0.58 = 0.42

P(at least one is read to) = 1 − P(none are read to)
= 1 − P(all five are not read to)
= 1 − (0.42)^5 = 0.987

43.

P(at least one diamond)
= 1 − P(no diamond)
= $1 - \frac{39}{52} \cdot \frac{38}{51} \cdot \frac{37}{50} \cdot \frac{36}{49} \cdot \frac{35}{48} = 1 - \frac{69,090,840}{311,875,200}$
= $\frac{242,784,360}{311,875,200} = \frac{7,411}{9,520}$

45.

a. P(no video games are rated mature) = 1 − 0.155 = 0.845
P(none of the five were rated mature) = (0.845)^5 = 0.4308

b. P(at least one of the five was rated mature) = 1 − 0.4308 = 0.5692

47.

P(at least one odd number) = 1 − P(no odd number)
= $1 - \left(\frac{1}{2}\right)^3 = 1 - \frac{1}{8} = 0.875$

The event is likely to occur since the probability is high.

49.

P(at least one 6) = $\frac{11}{36}$ = 0.306

51.

P(at least one will consider himself lucky) = 1 − P(no one will consider himself lucky)
= 1 − (0.88)^3 = 0.319

53.

No, because P(A ∩ B) = 0 and
P(A ∩ B) ≠ P(A) · P(B)

55.

Yes.

P(enroll) = 0.55

P(enroll | DW) > P(enroll) which indicates that DW has a positive effect on enrollment.

P(enroll | LP) = P(enroll) which indicates that LP has no effect on enrollment.

P(enroll | MH) < P(enroll) which indicates that MH has a low effect on enrollment.

Thus, all students should meet with DW.

57.

The Addition Rule states that
P(A or B) = P(A) + P(B) − P(A and B) and if A and B are mutually exclusive,
P(A and B) = 0.

Then 0.601 = 0.342 + 0.279 − P(A and B)
0.601 = 0.621 − P(A and B)
P(A and B) = 0.02
Therefore, events A and B are not mutually exclusive.

If A and B are independent,
P(A and B) = P(A) · P(B)
0.02 ≠ (0.342)(0.279)
Therefore, A and B are not independent.

P(A | B) = $\frac{P(B \text{ and } A)}{P(B)} = \frac{0.02}{0.279} = 0.072$

P(not B) = 1 − 0.279 = 0.721

59.

P(black | bag 1 or black | bag 2) = $\frac{2}{15}$
P(black | bag 1) + P(black | bag 2) = $\frac{2}{15}$
$\frac{1}{2} \cdot \frac{1}{10} + \frac{1}{2} \cdot \frac{1}{1+x} = \frac{2}{15}$

$\frac{1}{20} + \frac{1}{2+2x} = \frac{2}{15}$

$\frac{2x + 22}{20(2+2x)} = \frac{2}{15}$

Chapter 4 - Probability and Counting Rules

59. continued

$$30x + 330 = 80 + 80x$$
$$-50x = -250$$
$$x = 5$$

There are 5 white marbles in Bag #2.

EXERCISE SET 4-4

1.
$10^5 = 100,000$
$10 \cdot 9 \cdot 8 \cdot 7 \cdot 6 = 30,240$

3.
$7! = 7 \cdot 6 \cdot 5 \cdot 4 \cdot 3 \cdot 2 \cdot 1 = 5040$

5.
$10^5 = 100,000$
$10 \cdot 9 \cdot 8 \cdot 7 \cdot 6 = 30,240$

7.
$6^5 = 7776$

9.
$8 \cdot 3 \cdot 5 = 120$

11.
$\frac{12!}{(12-7)!} = 3,991,680$
$\frac{8!}{(8-3)!} \cdot 4! = 8064$

13.
a. $11! = 11 \cdot 10 \cdot 9 \cdot 8 \cdot 7 \cdot 6 \cdot 5 \cdot 4 \cdot 3 \cdot 2 \cdot 1$
$= 39,916,800$

b. $9! = 9 \cdot 8 \cdot 7 \cdot 6 \cdot 5 \cdot 4 \cdot 3 \cdot 2 \cdot 1$
$= 362,880$

c. $0! = 1$

d. $1! = 1$

13. continued

e. $_6P_4 = \frac{6!}{(6-4)!} = \frac{6 \cdot 5 \cdot 4 \cdot 3 \cdot 2 \cdot 1}{2 \cdot 1} = 360$

f. $_{12}P_8 = \frac{12!}{(12-8)!}$
$= \frac{12 \cdot 11 \cdot 10 \cdot 9 \cdot 8 \cdot 7 \cdot 6 \cdot 5 \cdot 4 \cdot 3 \cdot 2 \cdot 1}{4 \cdot 3 \cdot 2 \cdot 1} = 19,958,400$

g. $_7P_7 = \frac{7!}{(7-7)!} = \frac{7 \cdot 6 \cdot 5 \cdot 4 \cdot 3 \cdot 2 \cdot 1}{1} = 5040$

h. $_4P_0 = \frac{4!}{(4-0)!} = \frac{4 \cdot 3 \cdot 2 \cdot 1}{4 \cdot 3 \cdot 2 \cdot 1} = 1$

i. $_9P_2 = \frac{9!}{(9-2)!} = \frac{9 \cdot 8 \cdot 7 \cdot 6 \cdot 5 \cdot 4 \cdot 3 \cdot 2 \cdot 1}{7 \cdot 6 \cdot 5 \cdot 4 \cdot 3 \cdot 2 \cdot 1} = 72$

j. $_{11}P_3 = \frac{11!}{(11-3)!}$
$= \frac{11 \cdot 10 \cdot 9 \cdot 8 \cdot 7 \cdot 6 \cdot 5 \cdot 4 \cdot 3 \cdot 2 \cdot 1}{8 \cdot 7 \cdot 6 \cdot 5 \cdot 4 \cdot 3 \cdot 2 \cdot 1} = 990$

15.
$_4P_4 = \frac{4!}{(4-4)!} = \frac{4 \cdot 3 \cdot 2 \cdot 1}{0!} = 24$

17.
$_9P_3 = \frac{9!}{(9-3)!} = 504$

19.
$_7P_4 = \frac{7!}{(7-4)!} = \frac{7 \cdot 6 \cdot 5 \cdot 4 \cdot 3 \cdot 2 \cdot 1}{3 \cdot 2 \cdot 1} = 840$

21.
$_{10}P_6 = \frac{10!}{(10-6)!} = \frac{10 \cdot 9 \cdot 8 \cdot 7 \cdot 6 \cdot 5 \cdot 4 \cdot 3 \cdot 2 \cdot 1}{4 \cdot 3 \cdot 2 \cdot 1} = 151,200$

23.
$_{50}P_4 = \frac{50!}{(50-4)!} = \frac{50!}{46!} = 5,527,200$

25.
Same task: $_{12}C_4 = \frac{12!}{(12-4)!4!} = 495$

Different tasks: $_{12}P_4 = \frac{12!}{(12-4)!} = 11,880$

Chapter 4 - Probability and Counting Rules

27.
$\frac{7!}{3! \cdot 2! \cdot 2!} = 210$

29.
$\frac{9!}{4! \cdot 3! \cdot 2!} = 1260$

31.
$\frac{12!}{6! \cdot 3! \cdot 3!} = 18,480$

33.
a. $\frac{5!}{3! \, 2!} = 10$

b. $\frac{8!}{5! \, 3!} = 56$

c. $\frac{7!}{3! \, 4!} = 35$

d. $\frac{6!}{4! \, 2!} = 15$

e. $\frac{6!}{2! \, 4!} = 15$

35.
$_{50}C_5 = \frac{50!}{45! \, 5!} = 2,118,760$

37.
$_{12}C_4 = \frac{12!}{8! \, 4!} = 495$

39.
$_{10}C_3 \cdot _6C_2 = \frac{10!}{7! \, 3!} \cdot \frac{6!}{4! \, 2!} = 1800$

41.
$5 \cdot 5 \cdot 4 \cdot 4 \cdot 4 \cdot 4 = 6400$

43.
$_{12}C_4 = 495$

$_7C_2 \cdot _5C_2 = 21 \cdot 10 = 210$

$_7C_2 \cdot _5C_2 + _7C_3 \cdot _5C_1 + _7C_4$
$= 21 \cdot 10 + 35 \cdot 5 + 35$
$= 210 + 175 + 35 = 420$

45.
The possibilities are CVV or VCV or VVV.

Assuming the same vowel can't be used twice in a "word":
$7 \cdot 5 \cdot 4 + 5 \cdot 7 \cdot 4 + 5 \cdot 4 \cdot 3 = 340$

Assuming the same vowel can be used twice in a "word":
$7 \cdot 5 \cdot 5 + 5 \cdot 7 \cdot 5 + 5 \cdot 5 \cdot 5 = 475$

47.
The possibilities are 2 men and 2 women, 4 men and no women, or no men and 4 women.

$_6C_2 \cdot _4C_2 + _6C_4 \cdot _4C_0 + _6C_0 \cdot _4C_4 =$
$\frac{6!}{4! \cdot 2!} \cdot \frac{4!}{2! \cdot 2!} + \frac{6!}{2! \cdot 4!} \cdot \frac{4!}{4! \cdot 0!} + \frac{6!}{6! \cdot 0!} \cdot \frac{4!}{0! \cdot 4!} =$
$90 + 15 + 1 = 106$

49.
There are $_7C_2 = 21$ tiles with unequal numbers and 7 tiles with equal numbers. Thus, the total number of tiles is 28.

51.
$_{12}C_2 \cdot _8C_2 \cdot _6C_2 = \frac{12!}{10! \, 2!} \cdot \frac{8!}{6! \, 2!} \cdot \frac{6!}{4! \, 2!} = 27,720$

53.
$_{10}C_3 \cdot _6C_2 \cdot _5C_1 = \frac{10!}{7! \, 3!} \cdot \frac{6!}{4! \, 2!} \cdot \frac{5!}{4! \, 1!} = 9,000$

55.
$_{20}C_8 = \frac{20!}{(20-8)! \, 8!}$
$= \frac{20 \cdot 19 \cdot 18 \cdot 17 \cdot 16 \cdot 15 \cdot 14 \cdot 13 \cdot 12!}{12! \cdot 8 \cdot 7 \cdot 6 \cdot 5 \cdot 4 \cdot 3 \cdot 2 \cdot 1} = 125,970$

57.
$_9C_5 = \frac{9!}{4! \, 5!} = \frac{9 \cdot 8 \cdot 7 \cdot 6 \cdot 5!}{4 \cdot 3 \cdot 2 \cdot 1 \cdot 5!} = 126$

59.
$_{17}C_2 = \frac{17!}{(17-2)! \, 2!} = \frac{17 \cdot 16 \cdot 15!}{15! \cdot 2 \cdot 1} = 136$

Chapter 4 - Probability and Counting Rules

61.
$_{11}C_3 = \frac{11!}{8!\,3!} = \frac{11\cdot 10\cdot 9\cdot 8!}{8!\cdot 3\cdot 2\cdot 1} = 165$

63.
$_6C_3 \cdot _5C_2 = \frac{6!}{3!\,3!} \cdot \frac{5!}{3!\,2!} = 200$

65.
$_8P_3 = \frac{8!}{5!} = \frac{8\cdot 7\cdot 6\cdot 5\cdot 4\cdot 3\cdot 2\cdot 1}{5\cdot 4\cdot 3\cdot 2\cdot 1} = 336$

67.
$_4C_1 + _4C_2 + _4C_3 + _4C_4 =$
$\frac{4!}{3!\cdot 1!} + \frac{4!}{2!\cdot 2!} + \frac{4!}{1!\cdot 3!} + \frac{4!}{0!\cdot 4!} =$
$4 + 6 + 4 + 1 = 15$

69.
a. $2! \cdot 4! = 48$
b. 60 ways
Using a table, list the number of ways in each column and multiply:

B	C	3	2	1	=	6
B	3	C	2	1	=	6
B	3	2	C	1	=	6
B	3	2	1	C	=	6
3	B	C	2	1	=	6
3	B	2	C	1	=	6
3	B	2	1	C	=	6
3	2	B	C	1	=	6
3	2	B	1	C	=	6
3	2	1	B	C	=	6
						60

c. $5! - 2\cdot 4! = 72$

71.
$_{(x+2)}C_x = \frac{(x+2)!}{(x+2-x)!\cdot x!}$

$= \frac{(x+2)(x+1)(x)(x-1)\cdots(3)(2)(1)}{2!\cdot x!}$

$= \frac{(x+2)(x+1)x!}{2\cdot x!} = \frac{(x+2)(x+1)}{2}$

EXERCISE SET 4-5

1.
P(2 face cards) $= \frac{12}{52} \cdot \frac{11}{51} = \frac{11}{221}$

3.
a. There are $_5C_3$ ways of selecting 3 men and $_9C_3$ total ways to select 3 people;
hence, P(all men) $= \frac{_5C_3}{_9C_3} = \frac{10}{84} = \frac{5}{42}$.

b. There are $_4C_3$ ways of selecting 3 women;
hence, P(all women) $= \frac{_4C_3}{_9C_3} = \frac{4}{84} = \frac{1}{21}$.

c. There are $_5C_2$ ways of selecting 2 men and $_4C_1$ ways of selecting one woman;
hence, P(2 men and 1 woman) $= \frac{_5C_2\cdot _4C_1}{_9C_3}$
$= \frac{10}{21}$.

d. There are $_4C_2$ ways to select two women and $_5C_1$ ways of selecting one man; hence,
P(2 women and 1 man) $= \frac{_4C_2\cdot _5C_1}{_9C_3} = \frac{5}{14}$.

5.
a. P(both are men) $= \frac{_6C_2\cdot _7C_0}{_{13}C_2} = \frac{15}{78} = 0.192$

b. P(both are women) $= \frac{_6C_0\cdot _7C_2}{_{13}C_2} = \frac{21}{78} = 0.269$

c. P(one man and one woman) =
$\frac{_6C_1\cdot _7C_1}{_{13}C_2} = \frac{42}{78} = 0.538$

d. P(twins) $= \frac{1}{78} = 0.013$

7.
$\frac{_3C_2}{_{10}C_2} = \frac{3}{45} = \frac{1}{15}$

Chapter 4 - Probability and Counting Rules

9.

P(at least one U.S) = 1 − P(none are U.S)

$= 1 - \frac{_7C_0 \cdot _{13}C_5}{_{20}C_5}$

$= 1 - \frac{1287}{15,504} = 0.917$

P(at least two U.S) = 1 − P(none or one U.S)

$= 1 - \left(\frac{_7C_0 \cdot _{13}C_5 + _7C_1 \cdot _{13}C_4}{_{20}C_5}\right)$

$= 1 - \frac{6292}{15,504} = 0.594$

P(all five U.S) $= \frac{_7C_5 \cdot _{13}C_0}{_{20}C_5} = \frac{21}{15,504} = 0.001$

11.

a. P(red) $= \frac{_{11}C_2}{_{19}C_2} = \frac{55}{171} = 0.322$

b. P(black) $= \frac{_8C_2}{_{19}C_2} = \frac{28}{171} = 0.164$

c. P(unmatched) $= \frac{_{11}C_1 \cdot _8C_1}{_{19}C_2} = \frac{88}{171} = 0.515$

d. It probably got lost in the wash!

13.

There are $6^3 = 216$ ways of tossing three dice, and there are 10 ways of getting a sum of 6 such as (1, 1, 4), (1, 2, 3), (2, 2, 2), (1, 4, 1), etc. Hence, the probability of rolling a sum of 6 is $\frac{10}{216} = \frac{5}{108}$.

15.

There are $5! = 120$ ways to arrange 5 washers in a row and 2 ways to have them in correct order, small to large or large to small; hence, the probability is $\frac{2}{120} = \frac{1}{60}$.

17.

P(berries are produced) = P(1 or 2 males)

P(1 or 2 males) $= \frac{_8C_2 \cdot _4C_1}{_{12}C_3} + \frac{_8C_1 \cdot _4C_2}{_{12}C_3}$

$= 0.509 + 0.218 = 0.727$

REVIEW EXERCISES - CHAPTER 4

1.

a. $\frac{1}{8} = 0.125$ b. $\frac{3}{8} = 0.375$

c. $\frac{4}{8} = 0.50$

3.

a. P(not used for taxes) = P(virus or other)

P(virus or other) $= \frac{5}{10} + \frac{2}{10} = 0.7$

b. P(taxes or other use) $= \frac{3}{10} + \frac{2}{10} = 0.5$

5.

P(neither) = 1 − (either)

1 − P(either) = 1 − (0.32 + 0.41 - 0.06)

P(neither) = 0.33

7.

P(either backup or GPS)

$= 0.6 + 0.4 - 0.2 = 0.8$

P(neither backup nor GPS)

$= 1 - 0.8 = 0.2$

9.

P(either a lawnmower or a weed wacker) $= 0.7 + 0.5 - 0.3 = 0.9$

11.

P(enrolled in an online course) $= \frac{1}{6}$ or 0.167

a. P(all 5 took an online course) $= (\frac{1}{6})^5 = 0.0001$

b. P(none took an online course) $= (\frac{5}{6})^5 = 0.402$

c. P(at least one took an online course)

$= 1 - $ P(none took an online course)

$= 1 - (\frac{5}{6})^5 = 0.598$

Chapter 4 - Probability and Counting Rules

13.

a. $\frac{26}{52} \cdot \frac{25}{51} \cdot \frac{24}{50} = \frac{2}{17}$

b. $\frac{13}{52} \cdot \frac{12}{51} \cdot \frac{11}{50} = \frac{33}{2550} = \frac{11}{850}$

c. $\frac{4}{52} \cdot \frac{3}{51} \cdot \frac{2}{50} = \frac{1}{5525}$

15.
Total number of movie releases = 1384

a. P(European) = $\frac{834}{1384} = 0.603$

b. P(US) = $\frac{471}{1384} = 0.340$

c. P(German or French) = $\frac{316}{1384} + \frac{132}{1384}$
$= \frac{448}{1384}$ or 0.324

d. P(German | European)
$= \frac{P(\text{European and German})}{P(\text{European})} = \frac{\frac{316}{1384}}{\frac{834}{1384}} = 0.379$

17.

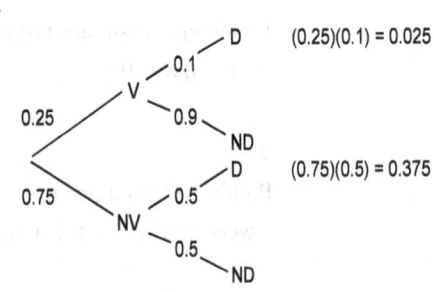

```
                    D   (0.25)(0.1) = 0.025
              0.1 ╱
           V ────
      0.25╱      0.9 ╲
         ╱              ND
        ╱           D   (0.75)(0.5) = 0.375
      0.75╲      0.5 ╱
           NV ───
              0.5 ╲
                    ND
```

P(disease) = 0.025 + 0.375 = 0.4

19.

P(NC | C) = $\frac{P(\text{NC and C})}{P(C)} = \frac{0.37}{0.73} = 0.507$

21.

$\frac{0.43}{0.75} = 0.573$ or 57.3%

23.

	<4 yrs HS	HS	College	Total
Smoker	6	14	19	39
Non-Smoker	<u>18</u>	<u>7</u>	<u>25</u>	<u>50</u>
Total	24	21	44	89

a. There are 44 college graduates and 19 of them smoke; hence, the probability is $\frac{19}{44}$.

b. There are 24 people who did not graduate from high school, 6 of whom do not smoke; hence, the probability is $\frac{6}{24} = \frac{1}{4}$.

25.
P(at least one household has a television set)
$= 1 - P(\text{none have a television set})$
$= 1 - (0.02)^4 = 0.99999984$

27.
If repetitions are allowed:
$26 \cdot 26 \cdot 10 \cdot 10 \cdot 10 = 676{,}000$

If repetitions are not allowed:
$_{26}P_2 \cdot {}_{10}P_3 = \frac{26 \cdot 25 \cdot 24!}{24!} \cdot \frac{10 \cdot 9 \cdot 8 \cdot 7!}{7!}$
$= 468{,}000$

If repetitions are allowed in the digits but not in the letters:
$10 \cdot 10 \cdot 10 \cdot {}_{26}P_2 = 650{,}000$

29.
$_5C_3 \cdot {}_7C_4 = \frac{5!}{2!\,3!} \cdot \frac{7!}{3!\,4!} = 10 \cdot 35 = 350$

31.
Although there are a total of 20 names, the names Ethan, Jacob and Noah appear on both lists. There are 17 different names to choose from.

$_{17}C_5 = 6188$ different ways to choose 5 names.

Chapter 4 - Probability and Counting Rules

33.
100!

35.
$_{12}C_4 = \frac{12!}{8!\,4!} = \frac{12\cdot11\cdot10\cdot9\cdot8!}{4\cdot3\cdot2\cdot1\cdot8!} = 495$

37.
$\frac{6!}{2!\,1!\,3!} = 60$

39.
$_{16}C_6 = \frac{16!}{10!\,6!} = \frac{16\cdot15\cdot14\cdot13\cdot12\cdot11\cdot10!}{10!\,6\cdot5\cdot4\cdot3\cdot2\cdot1} = 8008$

41.
Total catalog number of outcomes:
$26 \cdot 26 \cdot 10 \cdot 10 \cdot 10 = 676{,}000$
Total number of ways for ID followed by a number divisible by 5:
$26 \cdot 26 \cdot 10 \cdot 10 \cdot 2 = 135{,}200$
Hence, $P = \frac{135{,}200}{676{,}000} = 0.2$

43.
Total number of territories = 45
P(3 French or 3 UK or 3 US) $= \frac{_{16}C_3}{_{45}C_3} + \frac{_{15}C_3}{_{45}C_3} + \frac{_{14}C_3}{_{45}C_3}$

$= \frac{560}{14{,}190} + \frac{455}{14{,}190} + \frac{364}{14{,}190}$

$= \frac{1379}{14{,}190} = 0.097$

45.

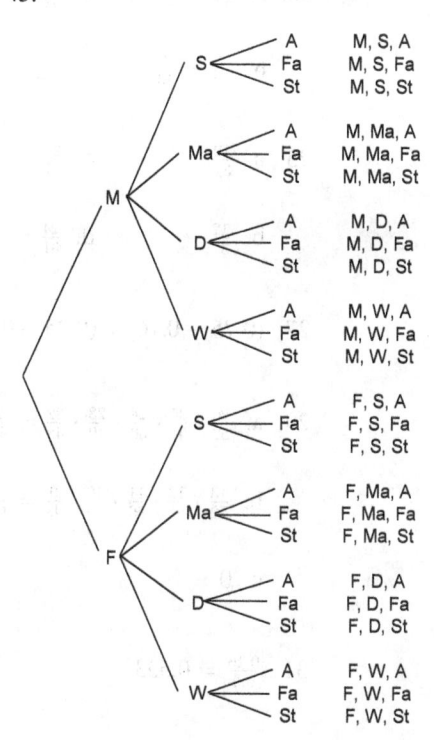

```
         ┌─ A   M, S, A
    ┌─ S ┼─ Fa  M, S, Fa
    │    └─ St  M, S, St
    │    ┌─ A   M, Ma, A
    ├─ Ma┼─ Fa  M, Ma, Fa
    │    └─ St  M, Ma, St
 M  │    ┌─ A   M, D, A
    ├─ D ┼─ Fa  M, D, Fa
    │    └─ St  M, D, St
    │    ┌─ A   M, W, A
    └─ W ┼─ Fa  M, W, Fa
         └─ St  M, W, St

         ┌─ A   F, S, A
    ┌─ S ┼─ Fa  F, S, Fa
    │    └─ St  F, S, St
    │    ┌─ A   F, Ma, A
    ├─ Ma┼─ Fa  F, Ma, Fa
    │    └─ St  F, Ma, St
 F  │    ┌─ A   F, D, A
    ├─ D ┼─ Fa  F, D, Fa
    │    └─ St  F, D, St
    │    ┌─ A   F, W, A
    └─ W ┼─ Fa  F, W, Fa
         └─ St  F, W, St
```

CHAPTER 4 QUIZ

1. False, subjective probability can be used when other types of probabilities cannot be found.

3. True

5. False, the probabilities can be different.

7. True

9. b

11. d

13. c

15. d

17. b

19. 0, 1

21. 1

Chapter 4 - Probability and Counting Rules

23. a. $\frac{4}{52} = \frac{1}{13}$ c. $\frac{16}{52} = \frac{4}{13}$

     b. $\frac{4}{52} = \frac{1}{13}$

25. a. $\frac{12}{31}$ c. $\frac{27}{31}$

     b. $\frac{12}{31}$ d. $\frac{24}{31}$

27. $(0.75 - 0.16) + (0.25 - 0.16) = 0.68$

29. a. $\frac{26}{52} \cdot \frac{25}{51} \cdot \frac{24}{50} \cdot \frac{23}{49} \cdot \frac{22}{48} = \frac{253}{9996}$

     b. $\frac{13}{52} \cdot \frac{12}{51} \cdot \frac{11}{50} \cdot \frac{10}{49} \cdot \frac{9}{48} = \frac{33}{66,640}$

     c. 0

31. $\frac{0.16}{0.3} = 0.533$

33. $\frac{0.028}{0.5} = 0.056$

35. $1 - (0.45)^6 = 0.992$

37. $1 - (0.15)^6 = 0.9999886$

39. 40,320

41. 1,188,137,600; 710,424,000

43. 33,554,432

45. $\frac{1}{4}$

47. $\frac{12}{55}$

49. 120,120

Chapter 5 - Discrete Probability Distributions

Note: Answers may vary due to rounding, TI-83's or computer programs.

EXERCISE SET 5-1

1. A random variable is a variable whose values are determined by chance. Examples will vary.

3. The number of commercials a radio station plays during each hour.
The number of times a student uses his or her calculator during a mathematics exam.
The number of leaves on a specific type of tree. (Answers will vary.)

5. For continuous variables, examples are length of home run, length of game, temperature at game time, pitcher's ERA, batting average., etc. For discrete variables, examples are number of hits, number of pitches, number of seats in each row, etc.

7. No; probabilities cannot be negative and the sum of the probabilities is not 1.

9. Yes

11. No; the sum of the probabilities is greater than 1.

13. Discrete

15. Continuous

17. Discrete

19.

X	0	1	2	3
P(X)	$\frac{4}{9}$	$\frac{2}{9}$	$\frac{2}{9}$	$\frac{1}{9}$

19. continued

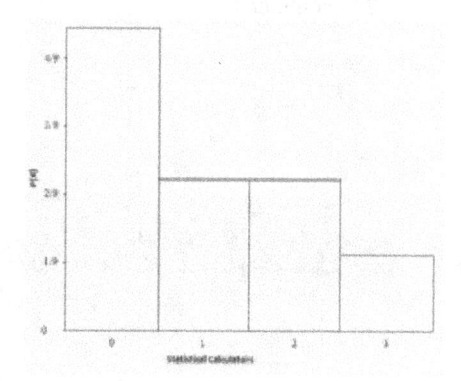

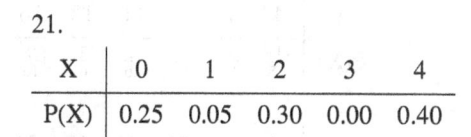

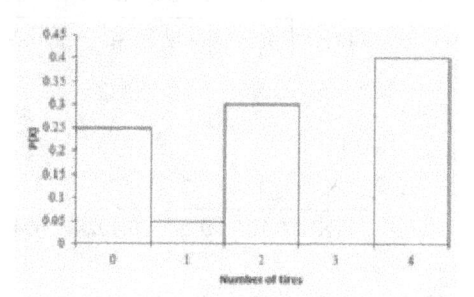

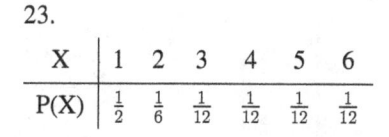

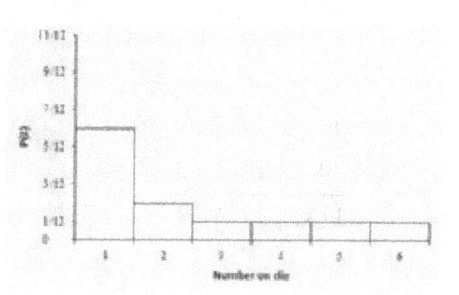

21.

X	0	1	2	3	4
P(X)	0.25	0.05	0.30	0.00	0.40

23.

X	1	2	3	4	5	6
P(X)	$\frac{1}{2}$	$\frac{1}{6}$	$\frac{1}{12}$	$\frac{1}{12}$	$\frac{1}{12}$	$\frac{1}{12}$

25.

X	2	3	4	5
P(X)	0.01	0.34	0.62	0.03

25. continued

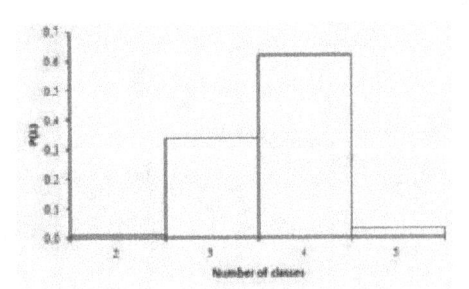

27.

X	4	7	9	11	13	16	18
P(X)	$\frac{1}{15}$	$\frac{1}{15}$	$\frac{1}{15}$	$\frac{1}{15}$	$\frac{1}{15}$	$\frac{2}{15}$	$\frac{1}{15}$

X	21	22	24	25	27	31	36
P(X)	$\frac{1}{15}$	$\frac{1}{15}$	$\frac{1}{15}$	$\frac{1}{15}$	$\frac{1}{15}$	$\frac{1}{15}$	$\frac{1}{15}$

29.

X	1	2	3	4	5
P(X)	0.124	0.297	0.402	0.094	0.083

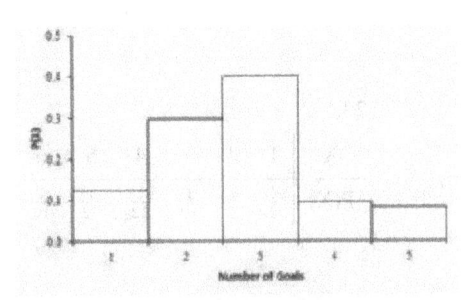

31.

X	1	2	3
P(X)	$\frac{1}{6}$	$\frac{1}{3}$	$\frac{1}{2}$

Yes.

33.

X	3	4	7
P(X)	$\frac{3}{6}$	$\frac{4}{6}$	$\frac{7}{6}$

No, the sum of the probabilities is greater than one and $P(7) = \frac{7}{6}$ which is also greater than 1.

35.

X	1	2	4
P(X)	$\frac{1}{7}$	$\frac{2}{7}$	$\frac{4}{7}$

Yes.

37.

X	1	2	3	4
P(X)	$\frac{3}{28}$	$\frac{6}{28}$	$\frac{12}{28}$	$\frac{7}{28}$

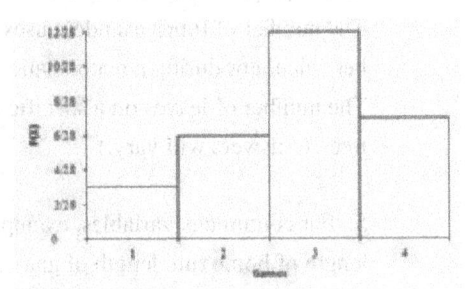

EXERCISE SET 5-2

1.

X	0	1	2	3	4
P(X)	0.31	0.42	0.21	0.04	0.02

$\mu = \sum X \cdot P(X) = 0(0.31) + 1(0.42) + 2(0.21) + 3(0.04) + 4(0.02) = 1.04$

$\sigma^2 = \sum X^2 \cdot P(X) - \mu^2 = [0^2(0.31) + 1^2(0.42) + 2^2(0.21) + 3^2(0.04) + 4^2(0.02)] - 1.04^2$

$\sigma^2 = 0.858$

$\sigma = \sqrt{0.858} = 0.926$

X	P(X)	X·P(X)	X²·P(X)
0	0.31	0	0
1	0.42	0.42	0.42
2	0.21	0.42	0.84
3	0.04	0.12	0.36
4	0.02	0.08	0.32
		$\mu = 1.04$	1.94

Chapter 5 - Discrete Probability Distributions

3.

$\mu = \sum X \cdot P(X) = 0(0.42) + 1(0.35) + 2(0.20) + 3(0.03) = 0.84$ or 0.8

$\sigma^2 = \sum X^2 \cdot P(X) - \mu^2 = [0^2(0.42) + 1^2(0.35) + 2^2(0.20) + 3^2(0.03)] - 0.84^2 = 0.71$ or 0.7

$\sigma = \sqrt{0.71} = 0.85$ or 0.9

X	P(X)	X·P(X)	X²·P(X)
0	0.42	0	0
1	0.35	0.35	0.35
2	0.20	0.40	0.80
3	0.03	0.09	0.27
		$\mu = 0.84$	1.42

5.

$\mu = \sum X \cdot P(X) = 6(0.30) + 7(0.40) + 8(0.25) + 9(0.05) = 7.05$

$\sigma^2 = \sum X^2 \cdot P(X) - \mu^2$
$= [6^2(0.30) + 7^2(0.40) + 8^2(0.25) + 9^2(0.05)] - 7.05^2$
$= 0.75$

$\sigma = \sqrt{0.75} = 0.86$

X	P(X)	X·P(X)	X²·P(X)
6	0.30	1.8	10.8
7	0.40	2.8	19.6
8	0.25	2.0	16.0
9	0.05	0.45	4.05
		$\mu = 7.05$	50.45

7.

$\mu = \sum X \cdot P(X) = 0(0.1) + 1(0.2) + 2(0.4) + 3(0.2) + 4(0.1) = 2.0$
$\mu = 2$

$\sigma^2 = \sum X^2 \cdot P(X) - \mu^2 = [0^2(0.1) + 1^2(0.2) + 2^2(0.4) + 3^2(0.2) + 4^2(0.1)] - 2.0^2 = 1.2$
$\sigma^2 = 1.2$

$\sigma = \sqrt{1.2} = 1.1$

X	P(X)	X·P(X)	X²·P(X)
0	0.1	0	0
1	0.2	0.2	0.2
2	0.4	0.8	1.6
3	0.2	0.6	1.8
4	0.1	0.4	1.6
		$\mu = 2.0$	5.2

9.

$\mu = \sum X \cdot P(X) = 1(0.27) + 2(0.46) + 3(0.21) + 4(0.05) + 5(0.01) = 2.07$ or 2.1

$\sigma^2 = \sum X^2 \cdot P(X) - \mu^2 = [1^2(0.27) + 2^2(0.46) + 3^2(0.21) + 4^2(0.05) + 5^2(0.01)] - 2.1^2 = 0.765$ or 0.80

$\sigma = \sqrt{0.80} = 0.89$ or 0.90

X	P(X)	X·P(X)	X²·P(X)
1	0.27	0.27	0.27
2	0.46	0.92	1.84
3	0.21	0.63	1.89
4	0.05	0.20	0.80
5	0.01	0.05	0.25
		$\mu = 2.1$	5.05

Chapter 5 - Discrete Probability Distributions

11.

Bag Value	X = Net Amt Won	P(X)
$1	−$1	$\frac{10}{20}$
$2	$0	$\frac{6}{20}$
$3	$1	$\frac{4}{20}$

$E(X) = \sum X \cdot P(X)$
$E(X) = [-\$1(\frac{10}{20}) + \$0(\frac{6}{20}) + \$1(\frac{4}{20})]$
$= -\$0.30$

13.

$E(X) = \sum X \cdot P(X) = \$5.00(\frac{1}{6}) = \$0.83$
He should pay about $0.83.

15.

$E(X) = \sum X \cdot P(X) = \$1000(\frac{1}{1000}) +$
$\$500(\frac{1}{1000}) + \$100(\frac{5}{1000}) - \$3.00$
$E(X) = -\$1.00$
Alternate Solution:
$E(X) = 997(\frac{1}{1000}) + 497(\frac{1}{1000})$
$+ 97(\frac{5}{1000}) - 3(\frac{993}{1000}) = -\$1.00$

17.

$E(X) = \sum X \cdot P(X)$
$= \$500(\frac{1}{1000}) - \$1.00$
$E(X) = -\$0.50$
Alternate Solution:
$E(X) = \$499(\frac{1}{1000}) - 1(\frac{999}{1000})$
$= -\$0.50$

There are 6 possibilities when a number with all different digits is boxed, $(3 \cdot 2 \cdot 1 = 6)$.
$E(X) = \$80(\frac{6}{1000}) - \$1.00$
$E(X) = \$0.48 - \$1.00 = -\$0.52$
Alternate Solution:
$E(X) = 79(\frac{6}{1000}) - 1(\frac{994}{1000}) = -\$0.52$

19.

a. $P(\text{red}) = \frac{18}{38}(\$1) + \frac{20}{38}(-\$1)$
$P(\text{red}) = -\$0.0526$

b. $P(\text{even}) = \frac{18}{38}(\$1) + \frac{20}{38}(-\$1)$
$P(\text{even}) = -\$0.0526$

c. $P(00) = \frac{1}{38}(\$35) + \frac{37}{38}(-\$1)$
$P(00) = -\$0.0526$

d. $P(\text{any single number}) =$
$\frac{1}{38}(\$35) + \frac{37}{38}(-\$1) = -\$0.0526$

e. $P(0 \text{ or } 00) =$
$\frac{2}{38}(\$17) + \frac{36}{38}(-\$1) = -\$0.0526$

21.

The expected value for a single die is 3.5, and since 3 die are rolled, the expected value is $3(3.5) = 10.5$

23.

Let $x = P(4)$. Then $\frac{2}{3}x = P(6)$.
$0.23 + 0.18 + x + \frac{2}{3}x + 0.015 = 1$
$0.425 + \frac{5}{3}x = 1$
$\frac{5}{3}x = 0.575$
$x = P(4) = 0.345$
Then $\frac{2}{3}x = P(6) = 0.23$

X	P(X)	X · P(X)	X² · P(X)
1	0.23	0.23	0.23
2	0.18	0.36	0.72
4	0.345	1.38	5.52
6	0.23	1.38	8.28
9	0.015	0.135	1.215
		$\mu = 3.485$	15.965

$\mu = 1(0.23) + 2(0.18) + 4(0.345) + 6(0.23)$
$+ 9(0.015) = 3.485$

$\sigma^2 = [1^2(0.23) + 2^2(0.18) + 4^2(0.345) +$
$6^2(0.23) + 9^2(0.015)] - 3.485^2 = 3.819$

$\sigma = \sqrt{3.819} = 1.954$

Chapter 5 - Discrete Probability Distributions

25.
Answers will vary.

27.
List the possible outcomes with the sums of the numbers on the balls:

(1,2) = 3	(4,1) = 5
(1,4) = 5	(4,2) = 6
(1,7) = 8	(4,7) = 11
(1,*) = 2	(4,*) = 8
(2,1) = 3	(7,1) = 8
(2,4) = 6	(7,2) = 9
(2,7) = 9	(7,4) = 11
(2,*) = 4	(7,*) = 14

(*,1) = 2
(*,2) = 4
(*,4) = 8
(*,7) = 14

Totals	f	P(X)
2	2	0.1
3	2	0.1
4	2	0.1
5	2	0.1
6	2	0.1
8	4	0.2
9	2	0.1
11	2	0.1
14	2	0.1

$\mu = 2(0.1) + 3(0.1) + 4(0.1) + 5(0.1) + 6(0.1) + 8(0.2) + 9(0.1) + 11(0.1) + 14(0.1) = 7$

$\sigma^2 = [2^2(0.1) + 3^2(0.1) + 4^2(0.1) + 5^2(0.1) + 6^2(0.1) + 8^2(0.2) + 9^2(0.1) + 11^2(0.1) + 14^2(0.1)] - 7^2 = 12.6$

$\sigma = \sqrt{12.6} = 3.55$

EXERCISE SET 5-3

1.
a. Yes
b. Yes
c. Yes
d. No, there are more than two outcomes.
e. No, there are more than two outcomes.

3.
a. 0.420
b. 0.346
c. 0.590
d. 0.251
e. 0.000

5.
a. $P(X) = \frac{n!}{(n-X)!\,X!} \cdot p^X \cdot q^{n-X}$

$P(X) = \frac{6!}{3! \cdot 3!} \cdot (0.03)^3 (0.97)^3$
$= 0.0005$

b. $P(X) = \frac{4!}{2! \cdot 2!} \cdot (0.18)^2 \cdot (0.82)^2$
$= 0.131$

c. $P(X) = \frac{5!}{2! \cdot 3!} = (0.63)^3 \cdot (0.37)^2$
$= 0.342$

7.
n = 10, p = 0.10
a. P(at least two) = 1 − P(none) − P(one)
$P(X \geq 2) = 1 - 0.349 - 0.387 = 0.264$

b.
P(X = 2 or 3) = 0.194 + 0.057 = 0.251

c.
P(X = 1) = 0.387

9.
a. n = 16, p = 0.20, X = 0
$P(X) = \frac{16!}{0!\,16!}(0.20)^0(0.80)^{16} = 0.028$

Chapter 5 - Discrete Probability Distributions

9. continued

b. $n = 16, P = 0.20, X = 8$

$P(X) = \frac{16!}{8!\,8!}(0.20)^8(0.80)^8 = 0.006$

c. $n = 16, p = 0.20, X = 4$

$P(X) = \frac{16!}{4!\,12!}(0.20)^4(0.80)^{12} = 0.200$

11.

$n = 9, p = \frac{3}{4} = 0.75, X = 5$

$P(X) = \frac{9!}{5!\,4!}(0.75)^5(0.25)^4 = 0.117$

13.

$n = 5, p = 0.40$

a. $X = 3, P(X) = 0.230$

b. $X = 0, 1, 2, 3,$ or 4 people
$P(X) = 0.078 + 0.259 + 0.346 + 0.230 + 0.077 = 0.99$

c. $n = 5, p = 0.40, X \geq 3$
$P(X) = 0.230 + 0.077 + 0.01 = 0.317$

d. $X = 0,$ or 1 person
$P(X) = 0.078 + 0.259 = 0.337$

15.

a. $n = 10, p = 0.53, X = 5$

$P(X) = \frac{10!}{5!5!}(0.53)^5(0.47)^5 = 0.242$

b. $n = 10, p = 0.47, X \geq 5$

$P(X) = \frac{10!}{5!5!}(0.47)^5(0.53)^5 +$

$\frac{10!}{6!4!}(0.47)^6(0.53)^4 + \frac{10!}{7!3!}(0.47)^7(0.53)^3 +$

$\frac{10!}{8!2!}(0.47)^8(0.53)^2 + \frac{10!}{9!1!}(0.47)^9(0.53)^1 +$

$\frac{10!}{10!0!}(0.47)^{10}(0.53)^0 = 0.548$

15. continued

c. $n = 10, p = 0.53, X < 5$

$P(X) = \frac{10!}{5!5!}(0.53)^5(0.47)^5 +$

$\frac{10!}{4!6!}(0.53)^4(0.47)^6 + \frac{10!}{3!7!}(0.53)^3(0.47)^7 +$

$\frac{10!}{2!8!}(0.53)^2(0.47)^8 + \frac{10!}{1!9!}(0.53)^1(0.47)^9$

$\frac{10!}{0!10!}(0.53^0)(0.47)^{10} = 0.306$

17.

a. $\mu = 100(0.75) = 75$
$\sigma^2 = 100(0.75)(0.25) = 18.75$ or 18.8
$\sigma = \sqrt{18.75} = 4.33$ or 4.3

b. $\mu = 300(0.3) = 90$
$\sigma^2 = 300(0.3)(0.7) = 63$
$\sigma = \sqrt{63} = 7.94$ or 7.9

c. $\mu = 20(0.5) = 10$
$\sigma^2 = 20(0.5)(0.5) = 5$
$\sigma = \sqrt{5} = 2.236$ or 2.2

d. $\mu = 10(0.8) = 8$
$\sigma^2 = 10(0.8)(0.2) = 1.6$
$\sigma = \sqrt{1.6} = 1.265$ or 1.3

19.

$n = 300, p = 0.25$
$\mu = 300(0.25) = 75$
$\sigma^2 = 300(0.75)(0.25) = 56.25$
$\sigma = \sqrt{56.25} = 7.5$

21.

If 13.1% are foreign-born, 86.9% are native Americans. In a sample of 60, one could expect $60(0.869) = 52.1$ or 52 approximately to be American-born.

Chapter 5 - Discrete Probability Distributions

21. continued

$n = 60, p = 0.131$
$\mu = 60(0.131) = 7.86$
$\sigma^2 = 60(0.131)(0.869) = 6.8$
$\sigma = \sqrt{6.8} = 2.6$

23.

$n = 1000, p = 0.21$
$\mu = 1000(0.21) = 210$
$\sigma^2 = 1000(0.21)(0.79) = 165.9$
$\sigma = \sqrt{165.9} = 12.9$

25.

$n = 18, p = 0.25, X = 5$
$P(X) = \frac{18!}{13!\,5!}(0.25)^5(0.75)^{13} = 0.199$

27.

$n = 10, p = \frac{1}{3}, X = 0, 1, 2, 3$
$P(X) = \frac{10!}{10!\,0!}(\frac{1}{3})^0(\frac{2}{3})^{10} + \frac{10!}{9!\,1!}(\frac{1}{3})^1(\frac{2}{3})^9$
$+ \frac{10!}{8!\,2!}(\frac{1}{3})^2(\frac{2}{3})^8 + \frac{10!}{7!\,3!}(\frac{1}{3})^3(\frac{2}{3})^7 = 0.559$

29.

$n = 20, p = 0.58, X = 14$
$P(X) = \frac{20!}{6!\,14!}(0.58)^{14}(0.42)^6 = 0.104$

31.

$n = 7, p = 0.14, X = 2 \text{ or } 3$
$P(X) = \frac{7!}{5!\,2!}(0.14)^2(0.86)^5 +$
$\frac{7!}{4!\,3!}(0.14)^3(0.86)^4 = 0.246$

33.

X	0	1	2	3
P(X)	0.125	0.375	0.375	0.125

35.

If $n = 3$, the outcomes and their corresponding probabilities are:

$X = 0; P(0) = \frac{3!}{3!\,0!}(p)^0(q)^3 = q^3$

$X = 1; P(1) = \frac{3!}{2!\,1!}(p)^1(q)^2 = 3pq^2$

$X = 2; P(2) = \frac{3!}{1!\,2!}(p)^2(q)^1 = 3p^2q$

$X = 3; P(3) = \frac{3!}{0!\,3!}(p)^3(q)^0 = p^3$

$\mu = \sum X \cdot P(X)$
$\mu = 0(q^3) + 1(3pq^2) + 2(3p^2q) + 3(p^3)$
$\mu = 3pq^2 + 6p^2q + 3p^3$
$\mu = 3p(q^2 + 2pq + p^2)$
$\mu = 3p(q+p)^2$
$\mu = 3p(1) = 3p$

EXERCISE SET 5-4

1.

a. $P(M) = \frac{6!}{3!\,2!\,1!}(0.5)^3(0.3)^2(0.2)^1$
$= 0.135$

b. $P(M) = \frac{5!}{1!\,2!\,2!}(0.3)^1(0.6)^2(0.1)^2$
$= 0.0324$

c. $P(M) = \frac{4!}{1!\,1!\,2!}(0.8)^1(0.1)^1(0.1)^2$
$= 0.0096$

3.

$P(M) =$
$\frac{12!}{2!\,2!\,2!\,2!\,2!\,2!}(0.13)^2(0.13)^2(0.14)^2(0.16)^2$
$(0.2)^2(0.24)^2 = 0.0025$

Chapter 5 - Discrete Probability Distributions

5.

$P(M) = \frac{10!}{3!3!2!2!}(0.5)^3(0.3)^3(0.15)^2(0.05)^2$
$= 0.0048$

7.
a. $P(6; 4) = 0.1042$
b. $P(2; 5) = 0.0842$
c. $P(7; 3) = 0.0216$

9.

$p = \frac{1}{20,000} = 0.00005$
$\lambda = n \cdot p = 80,000(0.00005) = 4$
a. $P(0; 4) = 0.0183$
b. $P(1; 4) = 0.0733$
c. $P(2; 4) = 0.1465$
d. $P(3 \text{ or more}; 4) = 1 - [P(0; 4) + P(1; 4) + P(2; 4)]$
$= 1 - (0.0183 + 0.0733 + 0.1465)$
$= 0.7619$

13.

$\lambda = 200(0.015) = 3$
$P(0; 3) = 0.0498$

15.

$P(5; 4) = 0.1563$

17.

P(one from each class)
$= \frac{{}_5C_1 \cdot {}_4C_1 \cdot {}_5C_1 \cdot {}_7C_1}{{}_{21}C_4} = \frac{700}{5985} = 0.117$

19.

P(three of them have a PhD)
$= \frac{{}_6C_3}{{}_{12}C_3} = 0.0909 \text{ or } 0.1$

21.

P(at least 1 defective) = 1 − P(0 defectives)
a = 6, b = 18, n = 3, X = 0

$P(0) = \frac{{}_6C_0 \cdot {}_{18}C_3}{{}_{24}C_3} = \frac{102}{253} = 0.403$

P(at least 1 defective) = 1 − 0.403 = 0.597

23.

P(yes) = 0.33 P(no) = 1 − 0.33 = 0.67
P("yes" fourth) = $(0.67)^3(0.33) = 0.099$

25.

P(bull's eye) = 0.4
P(no bull's eye) = 0.6
P(bull's eye on third shot)
$= (0.6)^2(0.4) = 0.144$

27.

$P(\text{club}) = p = \frac{13}{52} \text{ or } \frac{1}{4}$
$k = 3$
$\mu = \frac{3}{\frac{1}{4}} = 12$

29.

$P(\text{face card}) = \frac{12}{52} \text{ or } \frac{3}{13}$
$k = 4$
$\mu = \frac{4}{\frac{3}{13}} = 17.33 \text{ or } 18$

31.

$P(\text{prefer shower}) = p = \frac{4}{5} \text{ or } 0.8$
$q = 1 - 0.8 = 0.2$
$\mu = \frac{1}{0.8} = 1.25$
$\sigma = \sqrt{\frac{0.2}{(0.8)^2}} = 0.559$

33.

$p = \frac{1}{5} \text{ or } 0.2 \qquad q = \frac{4}{5} \text{ or } 0.8$
$\mu = \frac{1}{0.2} = 5$
$\sigma = \sqrt{\frac{0.8}{(0.2)^2}} = 4.472$

REVIEW EXERCISES - CHAPTER 5

1. Yes

3. No, the sum of the probabilities is greater than 1.

Chapter 5 - Discrete Probability Distributions

5.

X	0	1	2	3	4	5
P(X)	0.27	0.28	0.2	0.15	0.08	0.02

a. P(2 or 3 applications)
$= 0.2 + 0.15 = 0.35$

b.
$\mu = 0(0.27) + 1(0.28) + 2(0.2) + 3(0.15) + 4(0.08) + 5(0.02) = 1.55$

$\sigma^2 = [0^2(0.27) + 1^2(0.28) + 2^2(0.2) + 3^2(0.15) + 4^2(0.08) + 5^2(0.02)] - 1.55^2$
$= 1.8075$ or 1.808

$\sigma = \sqrt{1.8075} = 1.3444$

X	P(X)	X·P(X)	X²·P(X)
0	0.27	0	0
1	0.28	0.28	0.28
2	0.2	0.40	0.80
3	0.15	0.45	1.35
4	0.08	0.32	1.28
5	0.02	0.10	0.50
		μ = 1.55	4.21

7.

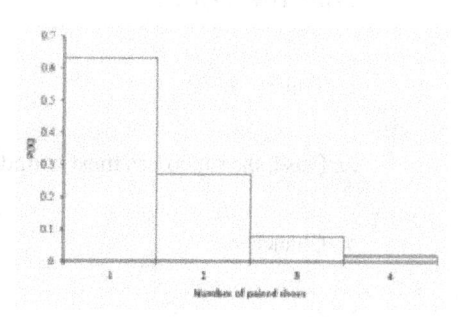

9.
$\mu = \sum X \cdot P(X) = 5(0.14) + 6(0.21) + 7(0.24) + 8(0.18) + 9(0.16) + 10(0.07)$
$= 7.22$

9. continued

$\sigma^2 = \sum X^2 \cdot P(X) - \mu^2 = [5^2(0.14) + 6^2(0.21) + 7^2(0.24) + 8^2(0.18) + 9^2(0.16) + 10^2(0.07)] - 7.22^2$
$= 2.1716$ or 2.2

$\sigma = \sqrt{2.1716} = 1.47$ or 1.5

X	P(X)	X·P(X)	X²·P(X)
5	0.14	0.7	3.5
6	0.21	1.26	7.56
7	0.24	1.68	11.76
8	0.18	1.44	11.52
9	0.16	1.44	12.96
10	0.07	0.7	7.00
		μ = 7.22	54.3

11.
$\mu = \sum X \cdot P(X) = 0(0.15) + 1(0.43) + 2(0.32) + 3(0.06) + 4(0.04) = 1.4$

$\sigma^2 = \sum X^2 \cdot P(X) - \mu^2 = [0^2(0.15) + 1^2(0.43) + 2^2(0.32) + 3^2(0.06) + 4^2(0.04)] - 1.4^2 = 0.90$

$\sigma = \sqrt{0.90} = 0.95$

X	P(X)	X·P(X)	X²·P(X)
0	0.15	0	0
1	0.43	0.43	0.43
2	0.32	0.64	1.28
3	0.06	0.18	0.54
4	0.04	0.16	0.64
		μ = 1.4	2.89

Two people at most should be employed.

Chapter 5 - Discrete Probability Distributions

13.
Let x = cost to play the game
$P(\text{ace}) = \frac{1}{13}$ $P(\text{face card}) = \frac{3}{13}$
$P(2 - 10) = \frac{9}{13}$
For a fair game, $E(X) = 0$.
$0 = -20(\frac{1}{13}) + 10(\frac{3}{13}) + 2(\frac{9}{13}) - x$
$x = \$2.15$

15.
$n = 12, p = 0.3$
a. $P(X = 8) = 0.008$
b. $P(X < 5) = 0.724$
c. $P(X \geq 10) = 0.0002$
d. $P(4 < x \leq 9) = 0.275$

17.
$\mu = n \cdot p = 250(0.58) = 145$
$\sigma^2 = n \cdot p \cdot q = 250(0.58)(0.42) = 60.9$
$\sigma = \sqrt{60.9} = 7.804$

19.
$n = 8, p = 0.25$
$P(X \leq 3) = \frac{8!}{8!\,0!}(0.25)^0(0.75)^8 +$
$\frac{8!}{7!\,1!}(0.25)^1(0.75)^7 + \frac{8!}{6!\,2!}(0.25)^2(0.75)^6 +$
$\frac{8!}{5!\,3!}(0.25)^3(0.75)^5 = 0.8862$ or 0.886

21.
$n = 20, p = 0.75, X = 16$
$P(16 \text{ have eaten pizza for breakfast}) =$
$\frac{20!}{4!\,16!}(0.75)^{16}(0.25)^4 = 0.1897$ or 0.190

23.
$P(M) = \frac{10!}{5!\,3!\,1!\,1!}(0.46)^5(0.41)^3(0.09)^1(0.04)^1$
$= 0.026$

25.
$P(M) = \frac{20!}{9!\,6!\,3!\,2!}(0.50)^9(0.28)^6(0.15)^3(0.07)^2$
$= 0.012$

27.
a. $P(6 \text{ or more}; 6) = 1 - P(5 \text{ or less}; 6)$
$= 1 - (0.0025 + 0.0149 + 0.0446 +$
$0.0892 + 0.1339 + 0.1606) = 0.5543$

b. $P(4 \text{ or more}; 6) = 1 - P(3 \text{ or less}; 6)$
$= 1 - (0.0025 + 0.0149 + 0.0446 +$
$0.0892) = 0.8488$

c. $P(5 \text{ or less}; 6) = P(0; 6) + ... + P(6; 6)$
$= 0.4457$

29.
$a = 13, b = 39, n = 5, X = 2$
$P(2) = \frac{_{13}C_2 \cdot _{39}C_3}{_{52}C_5} = \frac{9{,}139}{33{,}320} = 0.27$

31.
$P(1 \text{ vegetable \& 2 fruit}) =$
$\frac{_{10}C_0 \cdot _8C_1 \cdot _8C_2}{_{26}C_3} = \frac{224}{2600} = 0.086$

33.
$P(\text{face card on fourth draw}) =$
$(\frac{40}{52})^3(\frac{12}{52}) = 0.105$

CHAPTER 5 QUIZ

1. True

3. False, the outcomes must be independent.

5. Chance

7. One

9. c

11. No, the sum of the probabilities is greater than one.

13. Yes

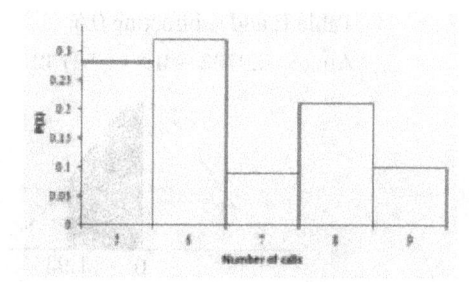

15.

[Histogram of P(x) vs Number of calls]

17.
$\mu = 0(0.10) + 1(0.23) + 2(0.31) + 3(0.27) + 4(0.09) = 2.02$ or 2
$\sigma^2 = [0^2(0.10) + 1^2(0.23) + 2^2(0.31) + 3^2(0.27) + 4^2(0.09)] - 2.02^2 = 1.3$
$\sigma = \sqrt{1.3} = 1.1$

19.
$\mu = 4(\frac{1}{6}) + 5(\frac{1}{6}) + 2(\frac{1}{6}) + 10(\frac{1}{6}) + 3(\frac{1}{6}) + 7(\frac{1}{6}) = 5.17$ or 5.2

21.
n = 20, p = 0.40, X = 5
P(5) = 0.124

23.
n = 300, p = 0.80
$\mu = 300(0.80) = 240$
$\sigma^2 = 300(0.80)(0.20) = 48$
$\sigma = \sqrt{48} = 6.9$

25.
P(M) $= \frac{30!}{15!\,8!\,5!\,2!}(0.5)^{15}(0.3)^8(0.15)^5(0.05)^2$
$= 0.008$

27.
P(M) $= \frac{12!}{5!\,4!\,3!}(0.45)^5(0.35)^4(0.2)^3$
$= 0.061$

29.
$\lambda = 8$
a. $P(X \geq 8; 8) = 0.1396 + \ldots + 0.0001$
$= 0.547$
b. $P(X \geq 3; 8) = 1 - P(0, 1, \text{ or } 2 \text{ calls})$
$= 1 - (0.0003 + 0.0027 + 0.0107)$
$= 1 - 0.0137 = 0.9863$
c. $P(X \leq 7; 8) = 0.0003 + \ldots + 0.1396$
$= 0.4529$

31.

a. $\frac{{}_6C_3 \cdot {}_8C_1}{{}_{14}C_4} = \frac{\frac{6!}{3!\,3!} \cdot \frac{8!}{7!\,1!}}{\frac{14!}{10!\,4!}} = 0.160$

b. $\frac{{}_6C_2 \cdot {}_8C_2}{{}_{14}C_4} = \frac{\frac{6!}{4!\,2!} \cdot \frac{8!}{6!\,2!}}{\frac{14!}{10!\,4!}} = 0.42$

c. $\frac{{}_6C_0 \cdot {}_8C_4}{{}_{14}C_4} = \frac{\frac{6!}{6!\,0!} \cdot \frac{8!}{4!\,4!}}{\frac{14!}{10!\,4!}} = 0.07$

33.
$P = (0.65)^9(0.35) = 0.007$

Chapter 6 - The Normal Distribution

Note to instructors: Graphs are not to scale and are intended to convey a general idea. Answers are generated using Table E. Answers generated using the TI calculator will vary slightly. Some TI calculator answers are shown.

EXERCISE SET 6-1

1.
The characteristics of the normal distribution are:

a. It is bell-shaped.

b. It is symmetric about the mean.

c. The mean, median, and mode are equal.

d. It is continuous.

e. It never touches the X-axis.

f. The area under the curve is equal to 1.

g. It is unimodal.

h. About 68% of the area lies within 1 standard deviation of the mean, about 95% within 2 standard deviations, and about 99.7% within 3 standard deviations of the mean.

3.
1 or 100%.

5.
68%, 95%, 99.7%

7.
The area is found by looking up $z = 1.07$ in Table E and subtracting 0.5.
Area $= 0.8577 - 0.5 = 0.3577$

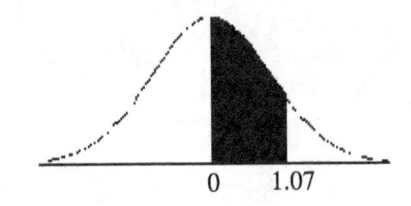

9.
The area is found by looking up $z = 1.93$ in Table E and subtracting 0.5.
Area $= 0.9732 - 0.5 = 0.4732$

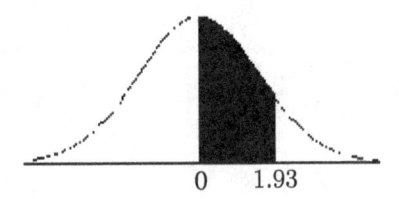

11.
The area is found by looking up $z = 0.37$ in Table E and subtracting from 1.
Area $= 1 - 0.6443 = 0.3557$

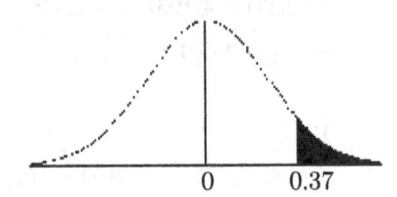

13.
The area is found by looking up $z = -1.87$ in Table E.
Area $= 0.0307$

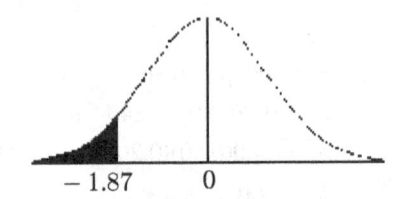

15.
The area is found by looking up the values 1.09 and 1.83 in Table E and subtracting the areas.
Area $= 0.9664 - 0.8621 = 0.1043$

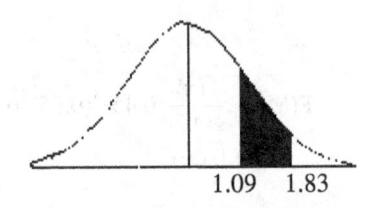

Chapter 6 - The Normal Distribution

17.
The area is found by looking up the values -1.46 and -1.77 in Table E and subtracting the areas.
Area $= 0.0721 - 0.0384 = 0.0337$

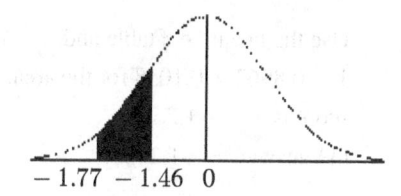

19.
The area is found by looking up the values -1.46 and -1.98 in Table E and subtracting the areas.
Area $= 0.0721 - 0.0239 = 0.0482$

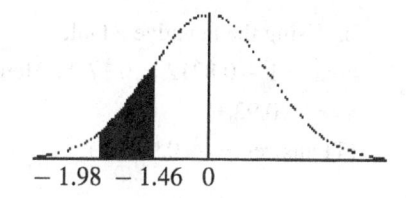

21.
The area is found by looking up 1.12 in Table E. Area $= 0.8686$

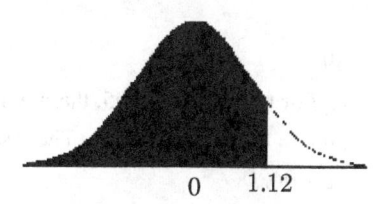

23.
The area is found by looking up -0.18 in Table E and subtracting it from 1.
$1 - 0.4286 = 0.5714$

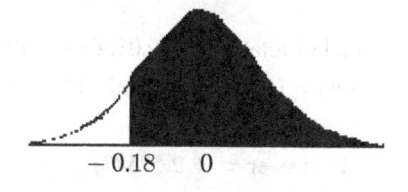

25.
For $z = -0.44$, the area is 0.3300. For $z = 1.92$, the area is $1 - 0.9726 = 0.0274$ Area $= 0.3300 + 0.0274 = 0.3574$

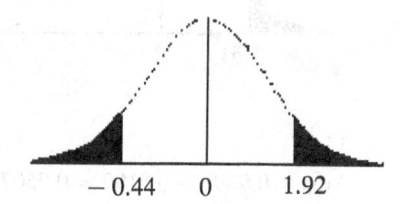

27.
Area $= 0.8289 - 0.5 = 0.3289$

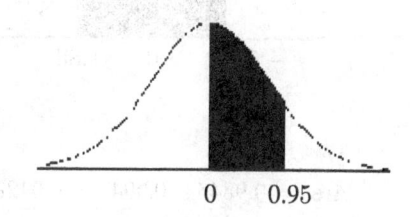

29.
Area $= 0.5 - 0.0838 = 0.4162$

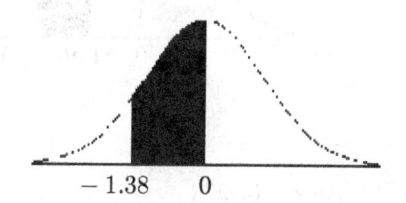

31.
Area $= 1 - 0.9901 = 0.0099$

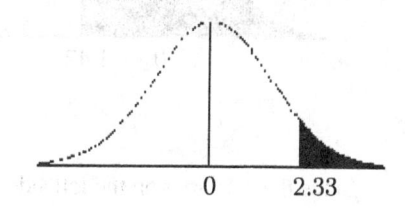

Chapter 6 - The Normal Distribution

33.
Area = 0.0655

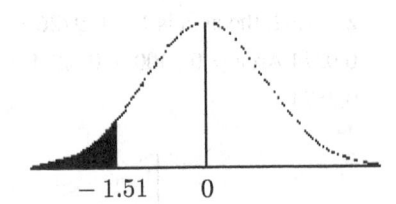

35.
Area = 0.9699 − 0.0192 = 0.9507

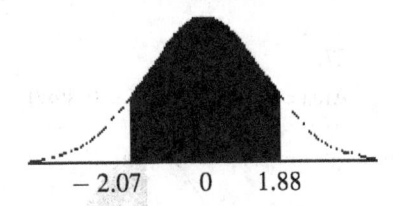

37.
Area = 0.9845 − 0.9947 = 0.0428

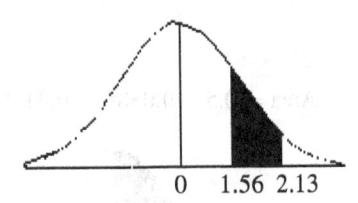

39.
Area = 0.9222

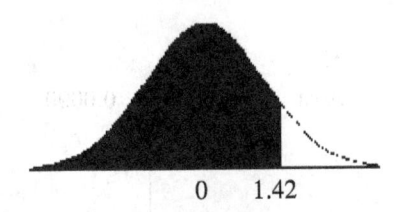

41.
Since the z score is on the left side of 0, use the negative z table. Areas in the negative z table are in the tail, so we will use $0.5 - 0.4175 = 0.0825$ as the area. The closest z score corresponding to an area of 0.0825 is $z = -1.39$.
(TI answer = −1.3885)

43.
$z = -2.08$, found by using the negative z table.
(TI answer = −2.0792)

45.
Use the negative z table and $1 - 0.8962 = 0.1038$ for the area. The z score is $z = -1.26$.
(TI answer = −1.2602)

47.
a. Using the negative z table, area $= 1 - 0.9887 = 0.0113$. Hence $z = -2.28$.
(TI answer = −2.2801)

b. Using the negative z table, area $= 1 - 0.8212 = 0.1788$. Hence $z = -0.92$.
(TI answer = −0.91995)

c. Using the negative z table, area $= 1 - 0.6064 = 0.3936$. Hence $z = -0.27$.
(TI answer = −0.26995)

49.
a. For total area = 0.05, there will be area = 0.025 in each tail. The z scores are ± 1.96.
(TI answer = ±1.95996)

b. For total area = 0.10, there will be area = 0.05 in each tail. The z scores are $z = \pm 1.645$.
(TI answer = ±1.64485)

c. For total area = 0.01, there will be area = 0.005 in each tail. The z scores are $z = \pm 2.58$.
(TI answer = ±2.57583)

Chapter 6 - The Normal Distribution

51.

$P(-1 < z < 1) = 0.8413 - 0.1587$
$= 0.6826$

$P(-2 < z < 2) = 0.9772 - 0.0228$
$= 0.9544$ (TI answer $= 0.9545$)

$P(-3 < z < 3) = 0.9987 - 0.0013$
$= 0.9974$ (TI answer $= 0.9973$)

They are very close.

53.
For $z = -1.2$, area $= 0.1151$
Area (left side) $= 0.5 - 0.1151 = 0.3849$
$0.8671 - 0.3849 = 0.4822$
Area (right side) $= 0.4822 + 0.5 = 0.9822$
For area $= 0.9822$, $z = 2.10$
Thus, $P(-1.2 < z < 2.10) = 0.8671$

55.
For $z = -0.5$, area $= 0.3085$
$0.3085 - 0.2345 = 0.074$
For area $= 0.074$, $z = -1.45$
Thus, $P(-1.45 < z < -0.5) = 0.2345$

For $z = -0.5$, area $= 0.3085$
$0.5 - 0.3085 = 0.1915$
$0.2345 - 0.1915 = 0.043$
$0.5 + 0.043 = 0.543$
For area $= 0.543$, $z = 0.11$
Thus, $P(-0.5 < z < 0.11) = 0.2345$

57.

$y = \dfrac{e^{-\frac{(X-0)^2}{2(1)^2}}}{1\sqrt{2\pi}} = \dfrac{e^{-\frac{x^2}{2}}}{\sqrt{2\pi}}$

59.
Since the area under the curve to the left of $z = 2.3$ and the area under the curve to the right of $z = -1.2$ are overlapping areas, this covers the entire area under the curve. Thus, the total area is 1.00.

EXERCISE SET 6-2

1.

$z = \dfrac{40 - 43.7}{1.6} = -2.31$

$P(z < -2.31) = 0.0104$ or 1.04%

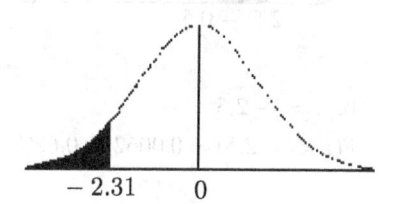

3.

a. $z = \dfrac{750{,}000 - 706{,}242}{52{,}145} = 0.84$

$P(z > 0.84) = 1 - 0.7995 = 0.2005$ or 20.05% (TI answer $= 0.2007$)

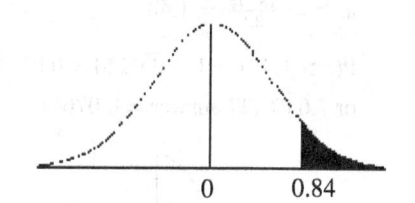

b. $z = \dfrac{600{,}000 - 706{,}242}{52{,}145} = -2.04$

$z = \dfrac{700{,}000 - 706{,}242}{52{,}145} = -0.12$

$P(-2.04 < z < -0.12) = 0.4522 - 0.0207$
$P = 0.4315$ or 43.15% (TI answer $= 0.4316$)

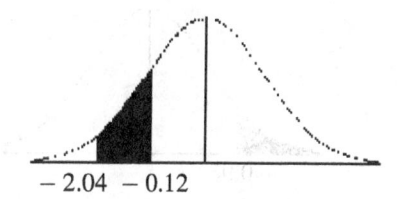

5.

a. $z = \dfrac{200 - 225}{10} = -2.5$

$z = \dfrac{220 - 225}{10} = -0.5$

$P(-2.5 < z < -0.5) =$
$0.3085 - 0.0062 = 0.3023$ or 30.23%

Chapter 6 - The Normal Distribution

5. continued

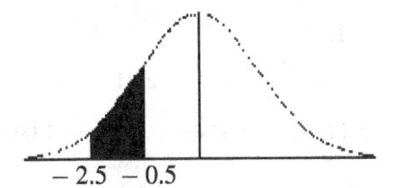

b. $z = -2.5$

$P(z < -2.5) = 0.0062$ or 0.62%

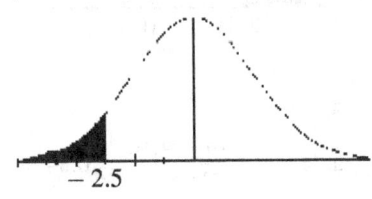

7.
a. $z = \frac{18-15}{2.1} = 1.43$

$P(z > 1.43) = 1 - 0.9236 = 0.0764$
or 7.64% (TI answer $= 0.0764$)

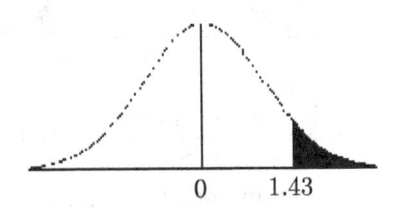

b. $z = \frac{13-15}{2.1} = -0.95$

$P(z < -0.95) = 0.1711$ or 17.11%
(TI answer $= 0.1711$)

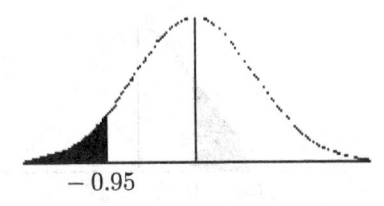

9.
For $x \geq 15,000$ miles:

$z = \frac{15,000-12,494}{1290} = 1.94$

$P(z > 1.94) = 1 - 0.9738 = 0.0262$
(TI answer $= 0.02603$)

9. continued

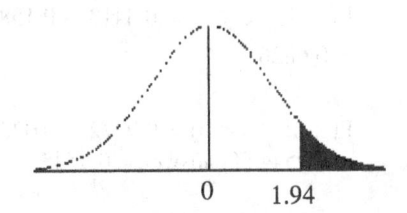

For $x < 8000$ miles:

$z = \frac{8000-12,494}{1290} = -3.48$

$P(z < -3.48) = 0.0003$
(TI answer $= 0.00025$)

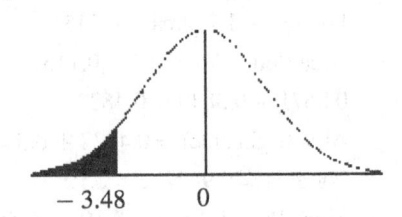

For $x < 6000$ miles:

$z = \frac{6000-12,494}{1290} = -5.03$

$P(z < -5.03) = 0.0001$

Maybe it would be good to know why it had only been driven less than 6000 miles.

11.
a. $z = \frac{1000-3262}{1100} = -2.06$

$P(z \geq -2.06) = 1 - 0.0197 = 0.9803$ or 98.03% (TI answer $= 0.9801$)

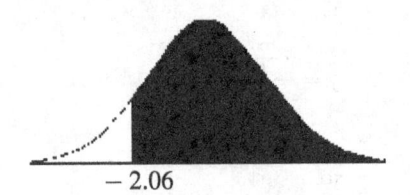

b. $z = \frac{4000-3262}{1100} = 0.67$

$P(z > 0.67) = 1 - 0.7486 = 0.2514$ or 25.14% (TI answer $= 0.2511$)

Chapter 6 - The Normal Distribution

11. continued

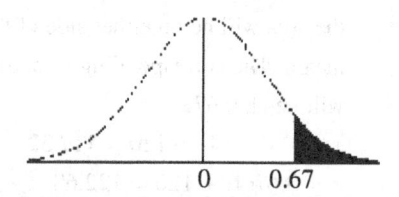

c. $z = \frac{3000 - 3262}{1100} = -0.24$

$P(-0.24 < z < 0.67) = 0.7486 - 0.4052 = 0.3434$ or 34.34%

(TI answer = 0.3430)

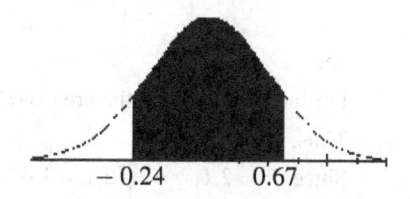

13.

a. $z = \frac{142 - 130}{5} = 2.4$

$P(z > 2.4) = 1 - 0.9918 = 0.0082$

(TI answer = 0.0082)

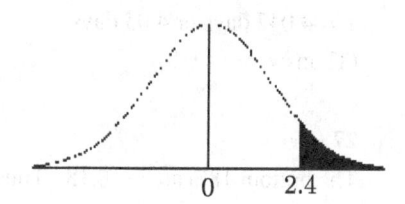

b. $z = \frac{125 - 130}{5} = -1$

$P(z < -1) = 0.1587$

(TI answer = 0.1587)

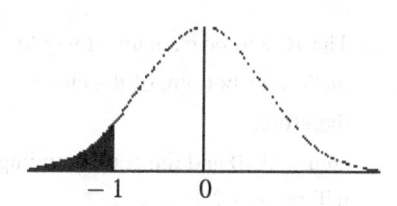

c. $z = \frac{136 - 130}{5} = 1.2$

$z = \frac{128 - 130}{5} = -0.4$

13. continued

$P(-0.4 < z < 1.2) = 0.8849 - 0.3446 = 0.5403$

(TI answer = 0.5403)

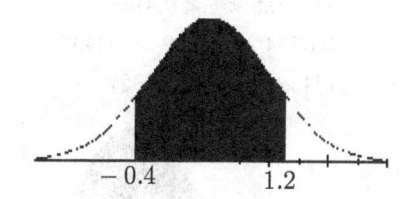

15.

a. $z = \frac{74 - 72}{3} = 0.67$

$z = \frac{68 - 72}{3} = -1.33$

$P(-1.33 < z < 0.67) = 0.7486 - 0.0918 = 0.6568$

(TI answer = 0.6568)

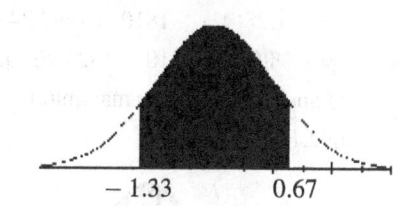

b.

$z = \frac{70 - 72}{3} = -0.67$

$P(z > -0.67) = 1 - 0.2514 = 0.7486$

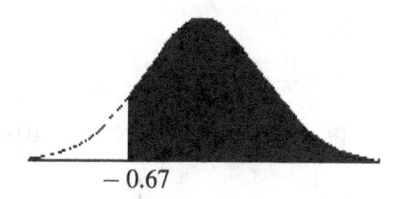

c. $z = \frac{75 - 72}{3} = 1$

$P(z < 1) = 0.8413$

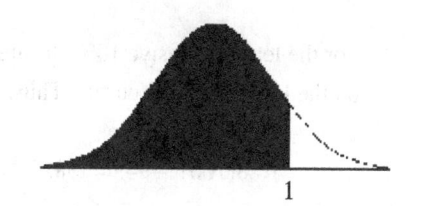

17.
$z = \frac{38-36}{5} = 0.4$

$z = \frac{32-36}{5} = -0.8$

$P(-0.8 < z < 0.4) =$
$0.6554 - 0.2119 = 0.4435$
(TI answer = 0.4435)

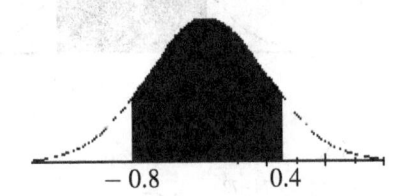

19.
The middle 80% means that 40% of the area will be on either side of the mean. The corresponding z scores will be ± 1.28.
$x = -1.28(92) + 1810 = 1692.24$ sq. ft.
$x = 1.28(92) + 1810 = 1927.76$ sq. ft.
(TI answers: 1927.90 maximum, 1692.10 minimum)

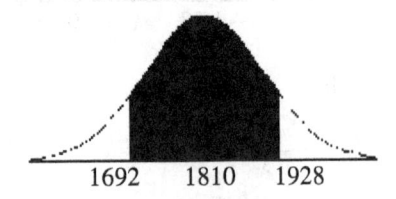

21.
$z = \frac{1200-949}{100} = 2.51$
$P(z > 2.51) = 1 - 0.9940 = 0.006$ or 0.6%

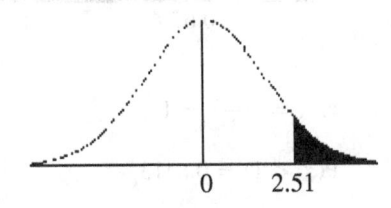

For the least expensive 10%, the area is 0.4 on the left side of the curve. Thus,
$z = -1.28$.
$x = -1.28(100) + 949 = \$821$

23.
The middle 50% means that 25% of the area will be on either side of the mean. The corresponding z scores will be ± 0.67.
$x = -0.67(4) + 120 = 117.32$
$x = 0.67(4) + 120 = 122.68$
(TI answer: $117.32 < \mu < 122.68$)

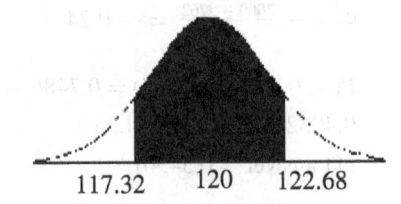

25.
For the longest 10%, the area is 0.90.
Thus, $z = 1.28$
Since $\sigma^2 = 2.1$, $\sigma = \sqrt{2.1} = 1.449$
$x = 1.28(1.449) + 4.8$
$x = 6.65$ or 6.7 days
(TI answer = 6.657)
For the shortest 30%, the area is 0.30.
Thus, $z = -0.52$.
$x = -0.52(1.449) + 4.8$
$x = 4.047$ days or 4.05 days
(TI answer = 4.040)

27.
The bottom 18% area is 0.18. Thus,
$z = -0.92$.
$x = -0.92(6256) + 24,596 = \$18,840.48$
(TI answer = \$18,869.48)

29.
The 10% to be exchanged would be at the left, or bottom, of the curve; therefore,
area = 0.10 and the corresponding z score will be -1.28.
$x = -1.28(5) + 25 = 18.6$ months.

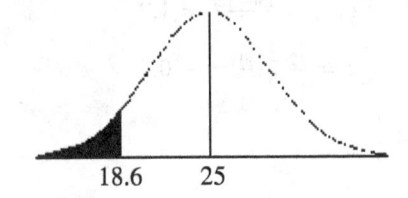

Chapter 6 - The Normal Distribution

31.
a. $\mu = 120 \quad \sigma = 20$
b. $\mu = 15 \quad \sigma = 2.5$
c. $\mu = 30 \quad \sigma = 5$

33.
For temperature of at least 85°, area is
$1 - 0.05 = 0.95$. Then $z = 1.645$
$85 = 1.645s + 73$
$s = 7.29°$

35.
For payments above $1255.94, area is
$1 - 0.25 = 0.75$. Then $z = 0.67$
$1255.94 = 0.67(120) + x$
$x = \$1175.54$

37.
Since $P(13.1 < x < 23.5) = 095$, the area on each side of the mean is 0.475. Thus, $z = \pm 1.96$.
$1.96 = \frac{23.5 - 18.3}{s}$
$s = 2.653$
$z = \frac{15 - 18.3}{2.653} = -1.24$
$P(z < -1.24) = 0.1075$

39.
Histogram:

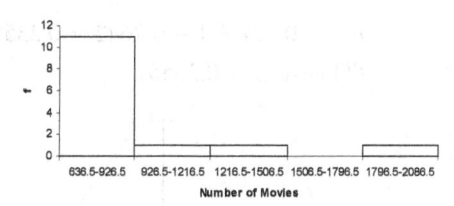

The histogram shows a positive skew.

$PI = \frac{3(970.2 - 853.5)}{376.5} = 0.93$

$IQR = Q_3 - Q_1 = 910 - 815 = 95$
$1.5(IQR) = 1.5(95) = 142.5$

$Q_1 - 142.5 = 672.5$
$Q_3 + 142.5 = 1052.5$

39. continued
There are several outliers.
Conclusion: The distribution is not normal.

41.
Histogram:

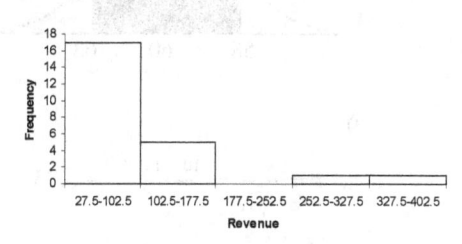

The histogram shows a positive skew.

$PI = \frac{3(90 - 59)}{89.598} = 1.04$

$IQR = Q_3 - Q_1 = 111 - 32 = 79$
$1.5(IQR) = 1.5(79) = 118.5$
$Q_1 - 118.5 = -86.5$
$Q_3 + 118.5 = 229.5$
There are two outliers.
Conclusion: The distribution is not normal.

43. Answers will vary.

EXERCISE SET 6-3

1.
The distribution is called the sampling distribution of sample means.

3.
The mean of the sample means is equal to the population mean.

5.
The distribution will be approximately normal when sample size is large.

7.
$z = \frac{\bar{X} - \mu}{\frac{\sigma}{\sqrt{n}}} = \frac{63 - 60}{\frac{8}{\sqrt{30}}} = 2.05$

$z = \frac{58 - 60}{\frac{8}{\sqrt{30}}} = -1.37$

Chapter 6 - The Normal Distribution

7. continued

$P(-1.37 < z < 2.05) = 0.9798 - 0.0853$
$\qquad = 0.8945$ or 89.45%

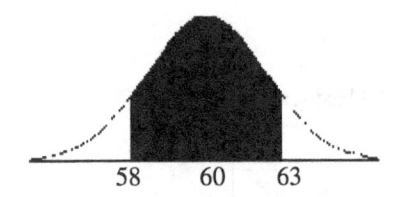

9.

a. $z = \dfrac{\overline{X}-\mu}{\frac{\sigma}{\sqrt{n}}} = \dfrac{10-12}{\frac{3.2}{\sqrt{36}}} = -3.75$

$P(z < -3.75) = 0.00009$
(TI answer $= 0.00009$)

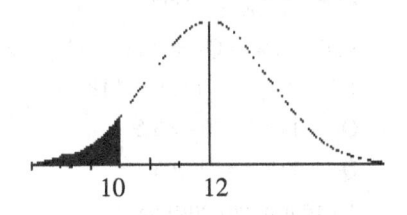

b. $z = \dfrac{\overline{X}-\mu}{\frac{\sigma}{\sqrt{n}}} = \dfrac{10-12}{\frac{3.2}{\sqrt{36}}} = -3.75$

$P(z > -3.75) = 1 - 0.00009 = 0.99991$
(TI answer $= 0.99991$)

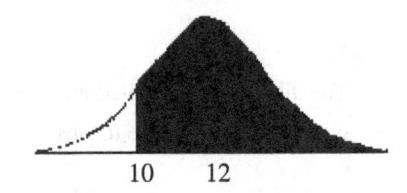

c. $z = \dfrac{\overline{X}-\mu}{\frac{\sigma}{\sqrt{n}}} = \dfrac{12-12}{\frac{3.2}{\sqrt{36}}} = 0$

$z = \dfrac{\overline{X}-\mu}{\frac{\sigma}{\sqrt{n}}} = \dfrac{11-12}{\frac{3.2}{\sqrt{36}}} = -1.88$

$P(-1.88 < z < 0) = 0.50 - 0.0301$
$\qquad = 0.4699$
(TI answer $= 0.4699$)

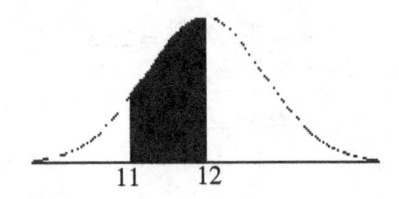

11.

$z = \dfrac{\overline{X}-\mu}{\frac{\sigma}{\sqrt{n}}} = \dfrac{37.5-36}{\frac{3.6}{\sqrt{35}}} = 2.47$

$z = \dfrac{\overline{X}-\mu}{\frac{\sigma}{\sqrt{n}}} = \dfrac{34-36}{\frac{3.6}{\sqrt{35}}} = -3.29$

$P(-3.29 < z < 2.47) = 0.9932 - 0.0005$
$\qquad = 0.9927$ or 99.27%
(TI answer $= 0.9927$)

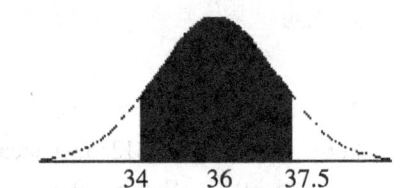

13.

$z = \dfrac{\overline{X}-\mu}{\frac{\sigma}{\sqrt{n}}} = \dfrac{8-8.61}{\frac{1.39}{\sqrt{50}}} = -3.10$

$P(z > -3.1) = 1 - 0.001 = 0.9990$
(TI answer $= 0.9990$)

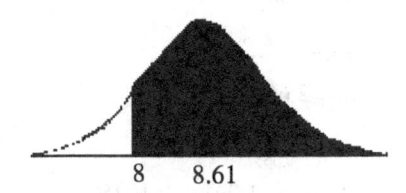

15.

$z = \dfrac{\overline{X}-\mu}{\sigma} = \dfrac{3000-2708}{405} = 0.72$

$P(z > 0.72) = 1 - 0.7642 = 0.2358$
(TI answer $= 0.2355$)

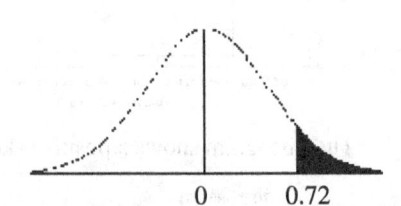

$z = \dfrac{\overline{X}-\mu}{\frac{\sigma}{\sqrt{n}}} = \dfrac{3000-2708}{\frac{405}{\sqrt{30}}} = 3.95$

$P(z > 3.95) = 1 - 0.9999 = 0.0001$
$P(z > 3.95) < 0.0001$
(TI answer $= 0.000039$)

Chapter 6 - The Normal Distribution

15. continued

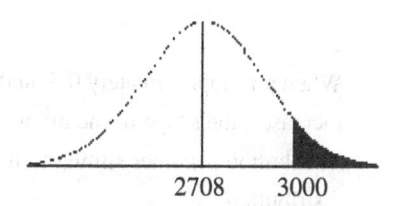

17.
$$z = \frac{\bar{X}-\mu}{\frac{\sigma}{\sqrt{n}}} = \frac{120-123}{\frac{21}{\sqrt{15}}} = -0.55$$

$$z = \frac{\bar{X}-\mu}{\frac{\sigma}{\sqrt{n}}} = \frac{126-123}{\frac{21}{\sqrt{15}}} = 0.55$$

$P(-0.55 < z < 0.55) = 0.7088 - 0.2912$
$= 0.4176$ or 41.76%
(TI answer $= 0.4199$)

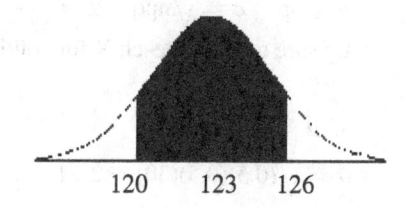

19.
$$z = \frac{\bar{X}-\mu}{\frac{\sigma}{\sqrt{n}}} = \frac{1980-2000}{\frac{187.5}{\sqrt{50}}} = -0.75$$

$$z = \frac{\bar{X}-\mu}{\frac{\sigma}{\sqrt{n}}} = \frac{1990-2000}{\frac{187.5}{\sqrt{50}}} = -0.38$$

$P(-0.75 < z < -0.38)$
$= 0.3520 - 0.2266 = 0.1254$
(TI answer $= 0.12769$)

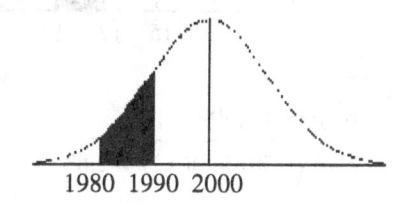

21.
a. $z = \frac{X-\mu}{\sigma} = \frac{8.2-8.9}{1.6} = -0.44$

$P(z < -0.44) = 0.33$ or 33%

21. continued

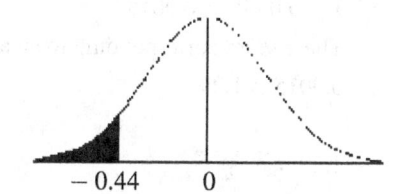

b. $z = \frac{8.2-8.9}{\frac{1.6}{\sqrt{10}}} = -1.38$

$P(z < -1.38) = 0.0838$ or 8.38%

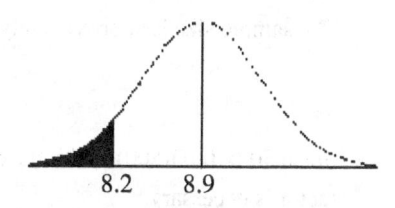

c. Yes, since the probability is slightly more than 30%.

d. Yes, but not as likely.

23.
a. $z = \frac{220-215}{15} = 0.33$

$P(z > 0.33) = 1 - 0.6293 = 0.3707$ or 37.07%
(TI answer $= 0.3694$)

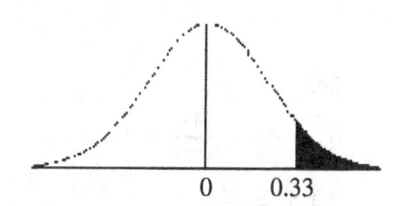

b. $z = \frac{220-215}{\frac{15}{\sqrt{25}}} = 1.67$

$P(z > 1.67) = 1 - 0.9525 = 0.0475$ or 4.75%
(TI answer $= 0.04779$)

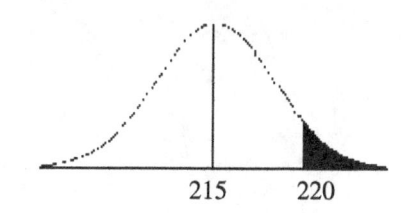

Chapter 6 - The Normal Distribution

25.

$1 - 0.0985 = 0.9015$

The z score corresponding to an area of 0.9015 is 1.29.

$1.29 = \frac{520-508}{\frac{72}{\sqrt{n}}}$

$1.29 = \frac{12\sqrt{n}}{72}$

$92.88 = 12\sqrt{n}$

$7.74 = \sqrt{n}$

$59.9 = n$

The sample size is approximately 60.

27.

Since $50 > 0.05(800)$ or 40, the correction factor is necessary.

It is $\sqrt{\frac{800-50}{800-1}} = 0.969$

$z = \frac{\overline{X}-\mu}{\frac{\sigma}{\sqrt{n}} \cdot \sqrt{\frac{N-n}{n-1}}} = \frac{83,500-82,000}{\frac{5000}{\sqrt{50}}(0.969)} = 2.19$

$P(z > 2.19) = 1 - 0.9857 = 0.0143$ or 1.43%

29.

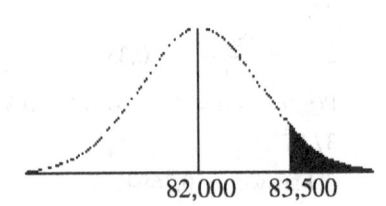

$\sigma_{\overline{X}} = \frac{\sigma}{\sqrt{n}} = \frac{15}{\sqrt{100}} = 1.5$

$2(1.5) = \frac{15}{\sqrt{n}}$

$3 \cdot \sqrt{n} = 15$

$\sqrt{n} = 5$

$n = 25$, the sample size necessary to double the standard error.

EXERCISE SET 6-4

1.

When p is approximately 0.5, and as n increases, the shape of the binomial distribution becomes similar to the normal distribution.

3.

The correction for continuity is necessary because the normal distribution is continuous and the binomial is discrete.

5.

For each problem use the following formulas:

$\mu = np \quad \sigma = \sqrt{npq} \quad z = \frac{\overline{X}-\mu}{\sigma}$

Be sure to correct each X for continuity.

a. $\mu = 0.5(30) = 15$

$\sigma = \sqrt{(0.5)(0.5)(30)} = 2.74$

$z = \frac{17.5-15}{2.74} = 0.91 \quad$ area $= 0.8186$

$z = \frac{18.5-15}{2.74} = 1.28 \quad$ area $= 0.8997$

$P(17.5 < X < 18.5) = 0.8997 - 0.8186$
$= 0.0811 = 8.11\%$

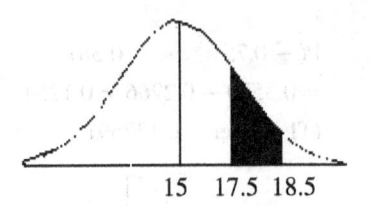

b. $\mu = 0.8(50) = 40$

$\sigma = \sqrt{(50)(0.8)(0.2)} = 2.83$

$z = \frac{43.5-40}{2.83} = 1.24 \quad$ area $= 0.8925$

$z = \frac{44.5-40}{2.83} = 1.59 \quad$ area $= 0.9441$

$P(43.5 < X < 44.5) = 0.9441 - 0.8925$
$= 0.0516$ or 5.16%

Chapter 6 - The Normal Distribution

5. continued

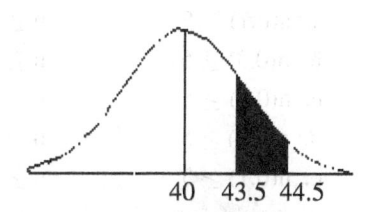

c. $\mu = 0.1(100) = 10$
$\sigma = \sqrt{(0.1)(0.9)(100)} = 3$

$z = \frac{11.5 - 10}{3} = 0.50 \qquad \text{area} = 0.6915$

$z = \frac{12.5 - 10}{3} = 0.83 \qquad \text{area} = 0.7967$

$P(11.5 < X < 12.5) = 0.7967 - 0.6915$
$\qquad\qquad\qquad\qquad = 0.1052$ or 10.52%

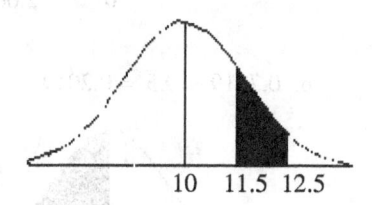

7.

a. $np = 20(0.50) = 10 \geq 5 \quad$ Yes
$\quad nq = 20(0.50) = 10 \geq 5$

b. $np = 10(0.60) = 6 \geq 5 \quad$ No
$\quad nq = 10(0.40) = 4 < 5$

c. $np = 40(0.90) = 36 \geq 5 \quad$ No
$\quad nq = 40(0.10) = 4 < 5$

9.

$\mu = 200(0.22) = 44$
$\sigma = \sqrt{(200)(0.22)(0.78)} = 5.8583$

$z = \frac{30.5 - 44}{5.8583} = -2.30 \qquad \text{area} = 0.0107$

$P(X > 30.5) = 1 - 0.0107 = 0.9893$

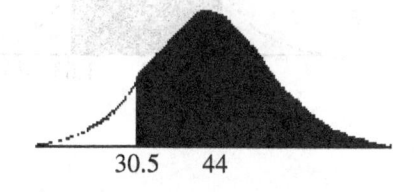

11.

$\mu = 120(0.659) = 79.08$
$\sigma = \sqrt{(120)(0.659)(0.341)} = 5.1929$

$z = \frac{64.5 - 79.08}{5.1929} = -2.81 \qquad \text{area} = 0.0025$

$z = \frac{85.5 - 79.08}{5.1929} = 1.24 \qquad \text{area} = 0.8925$

$P(64.5 < X < 85.5) = 0.8925 - 0.0025$
$P(64.5 \leq X \leq 85.5) = 0.8900$
(TI answer = 0.8893)

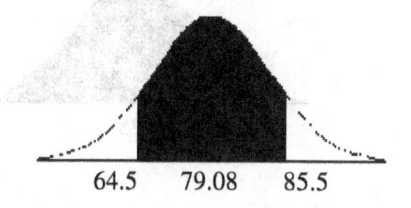

13.

$\mu = 60(0.76) = 45.6$
$\sigma = \sqrt{(60)(0.76)(0.24)} = 3.3082$

$z = \frac{48.5 - 45.6}{3.3082} = 0.88 \qquad \text{area} = 0.8106$

$z = \frac{47.5 - 45.6}{3.3082} = 0.57 \qquad \text{area} = 0.7157$

$P(47.5 < X < 48.5) = 0.8106 - 0.7157$
$P(47.5 < X < 48.5) = 0.0949$
(TI answer = 0.0949)

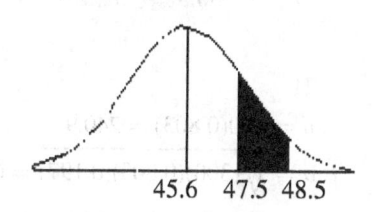

15.

$p = 0.22 \qquad \mu = 400(0.22) = 88$
$\sigma = \sqrt{(400)(0.22)(0.78)} = 8.2849$

$z = \frac{92.5 - 88}{8.2849} = 0.54$

$P(X \leq 92.5) = 0.7054$ or 70.54%

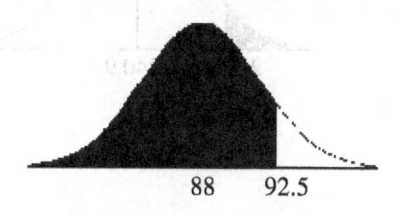

Chapter 6 - The Normal Distribution

17.

$\mu = 200(0.125) = 25$

$\sigma = \sqrt{(200)(0.125)(0.875)} = 4.6771$

$z = \dfrac{21.5 - 25}{4.6771} = -0.75$

$P(X \geq 21.5) = 1 - 0.2266 = 0.7734$

(TI answer = 0.7734)

Yes, it is very likely.

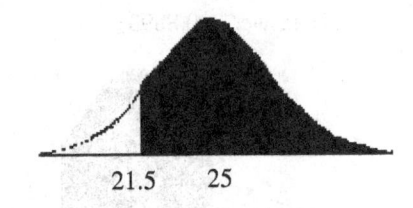

19.

$\mu = 200(0.261) = 52.2$

$\sigma = \sqrt{(200)(0.261)(0.739)} = 6.21$

$z = \dfrac{50.5 - 52.2}{6.21} = -0.27$

$P(X \leq 50.5) = 0.3936$

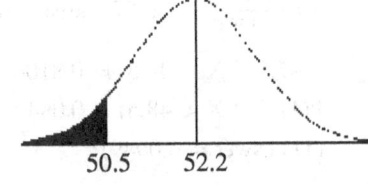

21.

$\mu = 300(0.803) = 240.9$

$\sigma = \sqrt{(300)(0.803)(0.197)} = 6.89$

$X < \frac{3}{4}(300)$ or $X < 225$

$z = \dfrac{224.5 - 240.9}{6.89} = -2.38$

$P(X < 224.5) = 0.0087$

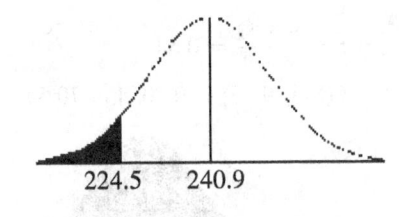

23.

a. $n(0.1) \geq 5$ $n \geq 50$

b. $n(0.3) \geq 5$ $n \geq 17$

c. $n(0.5) \geq 5$ $n \geq 10$

d. $n(0.2) \geq 5$ $n \geq 25$

e. $n(0.1) \geq 5$ $n \geq 50$

REVIEW EXERCISES - CHAPTER 6

1.

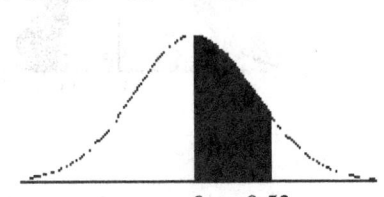

a. $0.9803 - 0.5 = 0.4803$

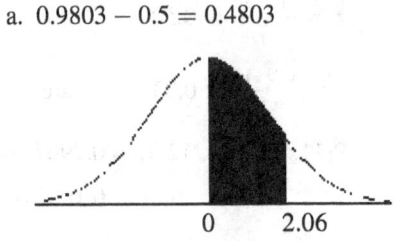

b. $0.7019 - 0.5 = 0.2019$

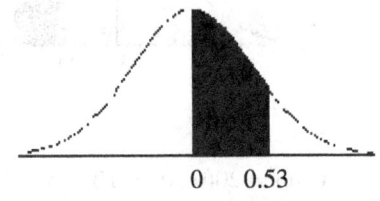

c. $0.9591 - 0.8962 = 0.0629$

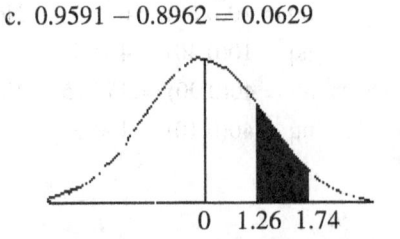

d. $0.9484 - 0.1539 = 0.7945$

(TI answer = 0.7945)

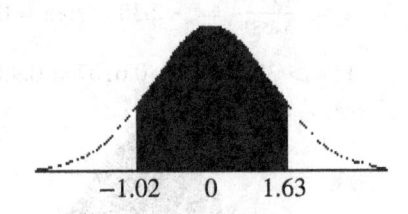

Chapter 6 - The Normal Distribution

1. continued

   e. $0.6879 - 0.4721 = 0.2158$

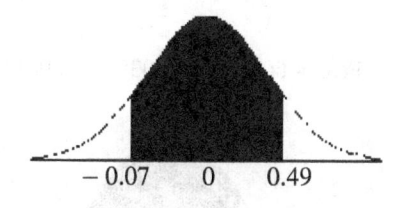

3.
   a. $0.9871 - 0.5 = 0.4871$

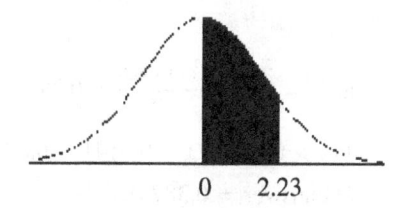

   b. $0.5 - 0.0401 = 0.4599$

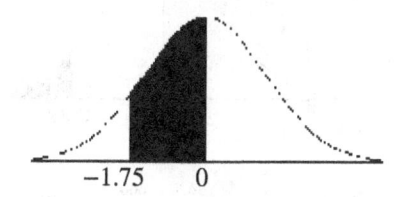

   c. $0.9535 - 0.0694 = 0.8841$

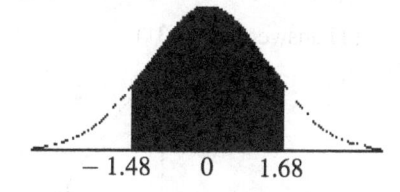

   d. $0.9616 - 0.8888 = 0.0728$

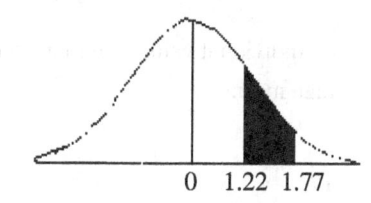

   e. $0.6255 - 0.0104 = 0.6151$

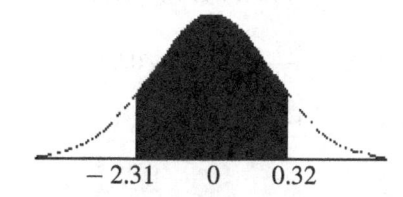

5. $z = \dfrac{\$6000 - \$5274}{\$600} = 1.21$

   $P(z > 1.21) = 1 - 0.8869 = 0.1131$

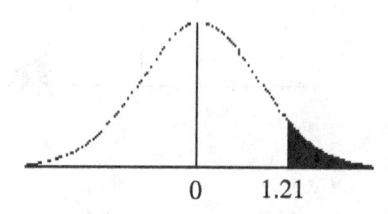

   For the middle 50%, 25% of the area is on each side of 0. Thus, $z = \pm 0.67$

   $x = 0.67(600) + 5274 = \$5676$
   $x = -0.67(600) + 5274 = \$4872$
   (TI answers: $4869.31 to $5678.69)

7.
   a. $z = \dfrac{476 - 476}{22} = 0$

   $z = \dfrac{500 - 476}{22} = 1.09$

   $P(0 < z < 1.09) = 0.8621 - 0.5 = 0.3621$ or 36.21%

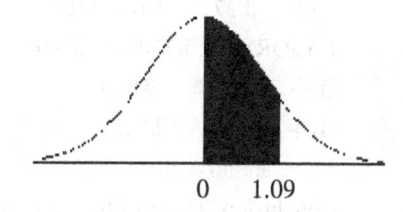

   b. $z = \dfrac{450 - 476}{22} = -1.18$

   $P(z < -1.18) = 0.1190$ or 11.9%

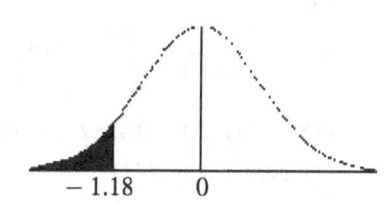

   c. $z = \dfrac{510 - 476}{22} = 1.55$

   $P(z > 1.55) = 1 - 0.9394 = 0.0606$ or 6.06%

Chapter 6 - The Normal Distribution

7. continued

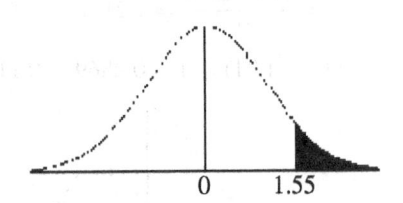

9.
For 15% costs, area = 0.85
$z = 1.04$
$X = 1.04(10.50) + 120 = \$130.92$

11.
Histogram:

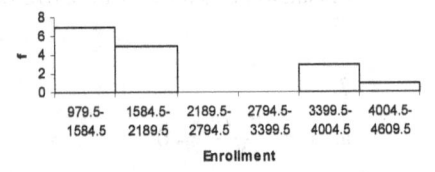

The histogram shows a positive skew.

$PI = \frac{3(2136.1 - 1755)}{1171.7} = 0.98$

$IQR = Q_3 - Q_1$

$IQR = 2827 - 1320 = 1507$

$1.5(IQR) = 1.5(1507) = 2260.5$

$Q_1 - 2260.5 = -940.5$

$Q_3 + 2260.5 = 5087.5$

There are no outliers.

Conclusion: The distribution is not normal.

13.

a. $z = \frac{X - \mu}{\frac{\sigma}{\sqrt{n}}} = \frac{27 - 25.7}{\frac{3.75}{\sqrt{40}}} = 2.19$

$P(\overline{X} > 27) = 1 - 0.9857 = 0.0143$
(TI answer = 0.0142)

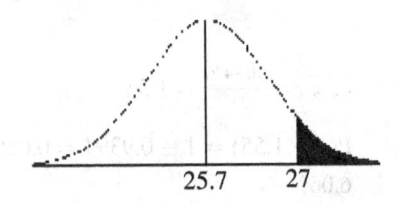

13. continued

b. $z = \frac{X - \mu}{\frac{\sigma}{\sqrt{n}}} = \frac{\$60 - \$61.50}{\frac{\$5.89}{\sqrt{50}}} = -1.80$

$P(\overline{X} > 60) = 1 - 0.0359 = 0.9641$

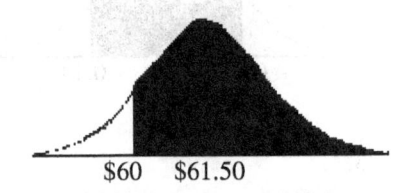

15.

a. $z = \frac{\overline{X} - \mu}{\sigma} = \frac{670 - 660}{35} = 0.29$

$P(z > 0.29) = 1 - 0.6141 = 0.3859$
(TI answer = 0.3875)

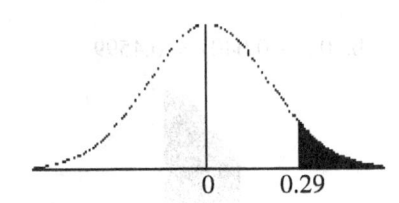

b. $z = \frac{\overline{X} - \mu}{\frac{\sigma}{\sqrt{n}}} = \frac{670 - 660}{\frac{35}{\sqrt{10}}} = 0.90$

$P(z > 0.90) = 1 - 0.8159 = 0.1841$
(TI answer = 0.1831)

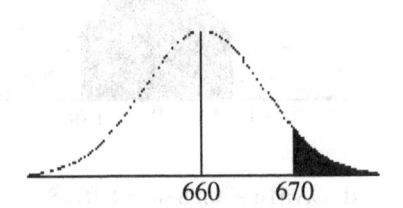

c. Individual values are more variable than means.

17.
$\mu = 120(0.173) = 20.76$

$\sigma = \sqrt{120(0.173)(0.827)} = 4.14$

$z = \frac{20.5 - 20.76}{4.14} = -0.06$

$z = \frac{34.5 - 20.76}{4.14} = 3.32$

Chapter 6 - The Normal Distribution

17. continued

P(20.5 < X < 34.5) = 0.9995 − 0.4761
   = 0.5234
(TI answer = 0.52456)

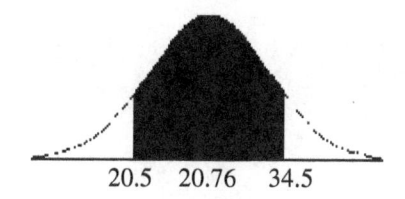

19.
For fewer than 10 holding multiple jobs:
$\mu = 150(0.053) = 7.95$
$\sigma = \sqrt{(150)(0.053)(0.947)} = 2.744$
$z = \frac{9.5 - 7.95}{2.74} = 0.56$

P(X < 9.5) = 0.7123
(TI answer = 0.7139)

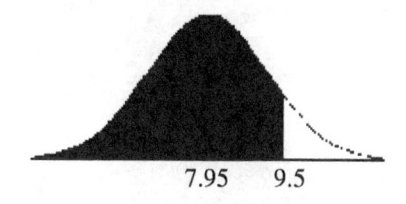

For more than 50 not holding multiple jobs: $\mu = 150(0.947) = 142.05$
$\sigma = \sqrt{150(0.947)(0.053)} = 2.744$
$z = \frac{50.5 - 142.05}{2.744} = -33.37$

P(X > 50.5) = 1 − 0.0001 = 0.9999
(TI answer = 0.9999)

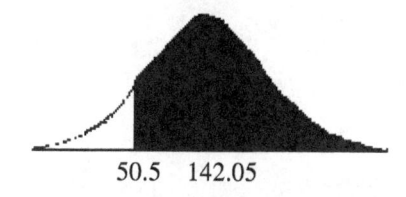

21.
$\mu = 200(0.37) = 74$
$\sigma = \sqrt{(200)(0.37)(0.63)} = 6.8279$

21. continued

$z = \frac{79.5 - 74}{6.8279} = 0.81$

P(X ≥ 79.5) = 1 − 0.7910 = 0.2090
or 20.90%

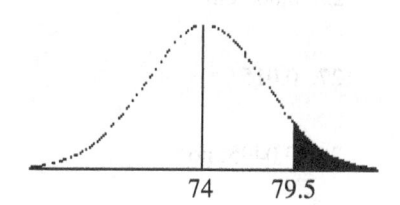

CHAPTER 6 QUIZ

1. False, the total area is equal to one.

3. True

5. False, the area is positive.

7. a

9. b

11. c

13. Sampling error

15. The standard error of the mean

17. 5%

19. The probabilities are:
   a. 0.4846   f. 0.0384
   b. 0.4693   g. 0.0089
   c. 0.9334   h. 0.9582
   d. 0.0188   i. 0.9788
   e. 0.7461   j. 0.8461

21. The probabilities are:
   a. 0.0668   c. 0.4649
   b. 0.0228   d. 0.0934

Chapter 6 - The Normal Distribution

23. The probabilities are:
    a. 0.0013   c. 0.0081
    b. 0.5   d. 0.5511

25. 8.804 cm

27. 0.015

29. 0.0495; no

31. 0.0614

33. The distribution is not normal.

Chapter 7 - Confidence Intervals and Sample Size

Note: Answers may vary due to rounding.

EXERCISE SET 7-1

1.

A point estimate of a parameter specifies a specific value such as $\mu = 87$, whereas an interval estimate specifies a range of values for the parameter such as $84 < \mu < 90$. The advantage of an interval estimate is that a specific confidence level (say 95%) can be selected, and one can be 95% confident that the parameter being estimated lies in the interval.

3.

The margin of error is the likely range of values above or below the statistic that may contain the parameter.

5.

A good estimator should be unbiased, consistent, and relatively efficient.

7.
a. 2.58   d. 1.65
b. 2.33   e. 1.88
c. 1.96

9.

For 95% confidence, $z_{\frac{\alpha}{2}} = 1.96$

$\overline{X} = 28.1 \quad \sigma = 4.7$

$28.1 - 1.96(\frac{4.7}{\sqrt{40}}) < \mu < 28.1 + 1.96(\frac{4.7}{\sqrt{40}})$

$28.1 - 1.5 < \mu < 28.1 + 1.5$

$26.6 < \mu < 29.6$

11.

a. $\overline{X} = 30$ is the point estimate for μ.

b. $\overline{X} - z_{\frac{\alpha}{2}}(\frac{\sigma}{\sqrt{n}}) < \mu < \overline{X} + z_{\frac{\alpha}{2}}(\frac{\sigma}{\sqrt{n}})$

$30 - (1.96)(\frac{4.2}{\sqrt{60}}) < \mu < 30 + (1.96)(\frac{4.2}{\sqrt{60}})$

$30 - 1.06 < \mu < 30 + 1.06$

$28.94 < \mu < 31.06$ or $28.9 < \mu < 31.1$

11. continued

c. $30 - (2.58)(\frac{4.2}{\sqrt{60}}) < \mu < 30 + (2.58)(\frac{4.2}{\sqrt{60}})$

$30 - 1.40 < \mu < 30 + 1.40$

$28.6 < \mu < 31.4$

d. The 99% confidence interval is larger because the confidence level is larger.

13.

For 92% confidence, $z_{\frac{\alpha}{2}} = 1.75$

$\overline{X} = 346.25 \quad \sigma = 165.1$

$346.25 - 1.75(\frac{165.1}{\sqrt{32}}) < \mu < 346.25 + 1.75(\frac{165.1}{\sqrt{32}})$

$346.25 - 51.08 < \mu < 346.25 + 51.08$

$295.2 < \mu < 397.3$

15.

$\overline{X} = 58.17 \quad \sigma = 6.46$

$58.17 - 1.96(\frac{6.46}{\sqrt{35}}) < \mu < 58.17 + 1.96(\frac{6.46}{\sqrt{35}})$

$58.17 - 2.14 < \mu < 58.17 + 2.14$

$56.1 < \mu < 60.3$

17.

$\overline{X} = 749 \quad \sigma = 32$

$749 - 1.96(\frac{32}{\sqrt{36}}) < \mu < 749 + 1.96(\frac{32}{\sqrt{36}})$

$738.5 < \mu < 759.5$ or $739 < \mu < 760$

803 gallons per year does not seem reasonable since it is outside this interval.

19.

$\overline{X} - z_{\frac{\alpha}{2}}(\frac{s}{\sqrt{n}}) < \mu < \overline{X} + z_{\frac{\alpha}{2}}(\frac{s}{\sqrt{n}})$

$61.2 - 1.96(\frac{7.9}{\sqrt{84}}) < \mu < 61.2 + 1.96(\frac{7.9}{\sqrt{84}})$

$61.2 - 1.69 < \mu < 61.2 + 1.69$

$59.5 < \mu < 62.9$

21.

$n = [\frac{z_{\frac{\alpha}{2}} \cdot \sigma}{E}]^2 = [\frac{(1.96)(7.5)}{2}]^2$

$n = (7.35)^2 = 54.0225$ or 55

Chapter 7 - Confidence Intervals and Sample Size

23.

$$n = \left[\frac{z_{\frac{\alpha}{2}} \sigma}{E}\right]^2 = \left[\frac{(1.65)(42)}{10}\right]^2$$

$n = (6.93)^2 = 48.02$ or 49 minutes

25.

$$n = \left[\frac{z_{\frac{\alpha}{2}} \sigma}{E}\right]^2 = \left[\frac{(2.58)(5.29)}{3}\right]^2$$

$n = (4.55)^2 = 20.7025$ or 21 days 2

EXERCISE SET 7-2

1.
The characteristics of the t-distribution are: It is bell-shaped, symmetrical about the mean, and never touches the x-axis. The mean, median, and mode are equal to 0 and are located at the center of the distribution. The variance is greater than 1. The t-distribution is a family of curves based on degrees of freedom. As sample size increases the t-distribution approaches the standard normal distribution.

3.
a. 2.898 where d. f. = 17
b. 2.074 where d. f. = 22
c. 2.624 where d. f. = 14
d. 1.833 where d. f. = 9
e. 2.093 where d. f. = 19

5.

$$\overline{X} - t_{\frac{\alpha}{2}}\left(\frac{s}{\sqrt{n}}\right) < \mu < \overline{X} + t_{\frac{\alpha}{2}}\left(\frac{s}{\sqrt{n}}\right)$$

$\overline{X} = 44.2 \qquad s = 2.6$

$$44.2 - (1.761)\left(\frac{2.6}{\sqrt{15}}\right) < \mu < 44.2 + (1.761)\left(\frac{2.6}{\sqrt{15}}\right)$$

$44.2 - 1.18 < \mu < 44.2 + 1.18$
$43 < \mu < 45$

7.
$\overline{X} = 33.4 \qquad s = 28.7$

$$\overline{X} - t_{\frac{\alpha}{2}}\left(\frac{s}{\sqrt{n}}\right) < \mu < \overline{X} + t_{\frac{\alpha}{2}}\left(\frac{s}{\sqrt{n}}\right)$$

$$33.4 - 1.746\left(\frac{28.7}{\sqrt{17}}\right) < \mu < 33.4 + 1.746\left(\frac{28.7}{\sqrt{17}}\right)$$

$33.4 - 12.2 < \mu < 33.4 + 12.2$
$21.2 < \mu < 45.6$

The point estimate is 33.4 and is close to the population mean of 32, which is within the 90% confidence interval. The mean may not be the best estimate since the data value 132 is large and possibly an outlier.

9.

$$\overline{X} - t_{\frac{\alpha}{2}}\left(\frac{s}{\sqrt{n}}\right) < \mu < \overline{X} + t_{\frac{\alpha}{2}}\left(\frac{s}{\sqrt{n}}\right)$$

$\overline{X} = 243.2 \qquad s = 23.8$

$$243.2 - 2.650\left(\frac{23.8}{\sqrt{14}}\right) < \mu < 243.2 + 2.650\left(\frac{23.8}{\sqrt{14}}\right)$$

$243.2 - 16.9 < \mu < 243.2 + 16.9$
$226.3 < \mu < 260.1$

11.

$$\overline{X} - t_{\frac{\alpha}{2}}\left(\frac{s}{\sqrt{n}}\right) < \mu < \overline{X} + t_{\frac{\alpha}{2}}\left(\frac{s}{\sqrt{n}}\right)$$

$$12{,}300 - 2.365\left(\frac{22}{\sqrt{8}}\right) < \mu < 12{,}300 + 2.365\left(\frac{22}{\sqrt{28}}\right)$$

$12{,}300 - 18 < \mu < 12{,}300 + 18$
$12{,}282 < \mu < 12{,}318$

The population mean for the weights of adult elephants is $12{,}282 < \mu < 12{,}318$.

13.

$$\overline{X} - t_{\frac{\alpha}{2}}\left(\frac{s}{\sqrt{n}}\right) < \mu < \overline{X} + t_{\frac{\alpha}{2}}\left(\frac{s}{\sqrt{n}}\right)$$

$$98 - 2.11\left(\frac{5.6}{\sqrt{18}}\right) < \mu < 98 + 2.11\left(\frac{5.6}{\sqrt{18}}\right)$$

$98 - 3 < \mu < 98 + 3$
$95 < \mu < 101$

Chapter 7 - Confidence Intervals and Sample Size

15.

$$\bar{X} - t_{\frac{\alpha}{2}}\left(\frac{s}{\sqrt{n}}\right) < \mu < \bar{X} + t_{\frac{\alpha}{2}}\left(\frac{s}{\sqrt{n}}\right)$$

$$109 - 3.106\left(\frac{4}{\sqrt{12}}\right) < \mu < 109 + 3.106\left(\frac{4}{\sqrt{12}}\right)$$

$$109 - 4 < \mu < 109 + 4$$

$$105 < \mu < 113$$

17.

$\bar{X} = 51.5 \quad s = 45.98$

$$\bar{X} - t_{\frac{\alpha}{2}}\left(\frac{s}{\sqrt{n}}\right) < \mu < \bar{X} + t_{\frac{\alpha}{2}}\left(\frac{s}{\sqrt{n}}\right)$$

$$51.5 - 1.746\left(\frac{45.98}{\sqrt{17}}\right) < \mu < 51.5 + 1.746\left(\frac{45.98}{\sqrt{17}}\right)$$

$$51.5 - 19.5 < \mu < 51.5 + 19.5$$

$$32.0 < \mu < 71.0$$

Assume a normal distribution.

19.

$$\bar{X} - t_{\frac{\alpha}{2}}\left(\frac{s}{\sqrt{n}}\right) < \mu < \bar{X} + t_{\frac{\alpha}{2}}\left(\frac{s}{\sqrt{n}}\right)$$

$\bar{X} = 40.494 \quad s = 31.9$

$$40.494 - 2.145\left(\frac{31.9}{\sqrt{15}}\right) < \mu < 40.494 + 2.145\left(\frac{31.9}{\sqrt{15}}\right)$$

$$40.494 - 17.667 < \mu < 40.494 + 17.667$$

$$22.827 < \mu < 58.161$$

21.

$\bar{X} = 2.175 \quad s = 0.585$

For $\mu > \bar{X} - t_{\frac{\alpha}{2}}\left(\frac{s}{\sqrt{n}}\right)$:

$$\mu > 2.175 - 1.729\left(\frac{0.585}{\sqrt{20}}\right)$$

$$\mu > 2.175 - 0.226$$

Thus, $\mu > \$1.95$ means that one can be 95% confident that the mean revenue is greater than $1.95.

For $\mu < \bar{X} + t_{\frac{\alpha}{2}}\left(\frac{s}{\sqrt{n}}\right)$:

$$\mu < 2.175 + 1.729\left(\frac{0.585}{\sqrt{20}}\right)$$

$$\mu < 2.175 + 0.226$$

Thus, $\mu < \$2.40$ means that one can be 95% confident that the mean revenue is less than $2.40.

EXERCISE SET 7-3

1.

a. $\hat{p} = \frac{40}{80} = 0.5 \qquad \hat{q} = \frac{40}{80} = 0.5$

b. $\hat{p} = \frac{90}{200} = 0.45 \qquad \hat{q} = \frac{110}{200} = 0.55$

c. $\hat{p} = \frac{60}{130} = 0.46 \qquad \hat{q} = \frac{70}{130} = 0.54$

d. $\hat{p} = 0.25 \qquad \hat{q} = 0.75$

e. $\hat{p} = 0.42 \qquad \hat{q} = 0.58$

3.

$\hat{p} = 0.33 \qquad \hat{q} = 0.67$

$$\hat{p} - (z_{\frac{\alpha}{2}})\sqrt{\frac{\hat{p}\hat{q}}{n}} < p < \hat{p} + (z_{\frac{\alpha}{2}})\sqrt{\frac{\hat{p}\hat{q}}{n}}$$

$$0.33 - (1.96)\sqrt{\frac{(0.33)(0.67)}{1000}} < p <$$
$$0.33 + (1.96)\sqrt{\frac{(0.33)(0.67)}{1000}}$$

$$0.33 - 0.029 < p < 0.33 + 0.029$$

$$0.301 < p < 0.359$$

5.

$\hat{p} = 0.68$

$\hat{q} = 1 - 0.68 = 0.32$

$$\hat{p} - (z_{\frac{\alpha}{2}})\sqrt{\frac{\hat{p}\hat{q}}{n}} < p < \hat{p} + (z_{\frac{\alpha}{2}})\sqrt{\frac{\hat{p}\hat{q}}{n}}$$

$$0.68 - 1.65\sqrt{\frac{(0.68)(0.32)}{100}} < p < 0.68 + 1.65\sqrt{\frac{(0.68)(0.32)}{100}}$$

$$0.68 - 0.077 < p < 0.68 + 0.077$$

0.603 or $60.3\% < p < 0.757$ or 75.7%

(TI answer: $0.603 < p < 0.757$)

7.

$\hat{p} = 0.84 \qquad \hat{q} = 0.16$

$$\hat{p} - (z_{\frac{\alpha}{2}})\sqrt{\frac{\hat{p}\hat{q}}{n}} < p < \hat{p} + (z_{\frac{\alpha}{2}})\sqrt{\frac{\hat{p}\hat{q}}{n}}$$

$$0.84 - 1.65\sqrt{\frac{(0.84)(0.16)}{200}} < p <$$
$$0.84 + 1.65\sqrt{\frac{(0.84)(0.16)}{200}}$$

$$0.84 - 0.043 < p < 0.84 + 0.043$$

$$0.797 < p < 0.883$$

9.
$\hat{p} = 0.65 \qquad \hat{q} = 0.36$

$\hat{p} - (z_{\frac{\alpha}{2}})\sqrt{\frac{\hat{p}\hat{q}}{n}} < p < \hat{p} + (z_{\frac{\alpha}{2}})\sqrt{\frac{\hat{p}\hat{q}}{n}}$

$0.65 - 1.96\sqrt{\frac{(0.65)(0.35)}{300}} < p <$

$\qquad 0.65 + 1.96\sqrt{\frac{(0.65)(0.35)}{300}}$

$0.65 - 0.054 < p < 0.65 + 0.054$

$0.596 < p < 0.704$

11.
$\hat{p} = 0.68 \qquad \hat{q} = 0.32$

$\hat{p} - (z_{\frac{\alpha}{2}})\sqrt{\frac{\hat{p}\hat{q}}{n}} < p < \hat{p} + (z_{\frac{\alpha}{2}})\sqrt{\frac{\hat{p}\hat{q}}{n}}$

$0.68 - 2.58\sqrt{\frac{(0.68)(0.32)}{50}} < p <$

$\qquad 0.68 + 2.58\sqrt{\frac{(0.68)(0.32)}{50}}$

$0.68 - 0.170 < p < 0.68 + 0.170$

$0.510 < p < 0.850$

13.
$\hat{p} = 0.86 \qquad \hat{q} = 0.14$

$\hat{p} - (z_{\frac{\alpha}{2}})\sqrt{\frac{\hat{p}\hat{q}}{n}} < p < \hat{p} + (z_{\frac{\alpha}{2}})\sqrt{\frac{\hat{p}\hat{q}}{n}}$

$0.86 - 2.58\sqrt{\frac{(0.86)(0.14)}{349}} < p <$

$\qquad 0.86 + 2.58\sqrt{\frac{(0.86)(0.14)}{349}}$

$0.86 - 0.048 < p < 0.86 + 0.048$

$0.812 < p < 0.908$

15.
$\hat{p} = \frac{40}{200} = 0.2 \qquad \hat{q} = 0.8$

$n = \hat{p}\hat{q}\left[\frac{z_{\frac{\alpha}{2}}}{E}\right]^2 = (0.2)(0.8)\left[\frac{1.96}{0.04}\right]^2$

$n = 384.16$ or 385

If no estimate of the sample proportion is available, use $\hat{p} = 0.5$:

$\hat{p} = 0.5 \quad \hat{q} = 0.5$

$n = \hat{p}\hat{q}\left[\frac{z_{\frac{\alpha}{2}}}{E}\right]^2 = (0.5)(0.5)\left[\frac{1.96}{0.04}\right]^2$

$n = 600.25$ or 601

17.
a. $\hat{p} = 0.25 \qquad \hat{q} = 0.75$

$n = \hat{p}\hat{q}\left[\frac{z_{\frac{\alpha}{2}}}{E}\right] = (0.25)(0.75)\left[\frac{1.96}{0.03}\right]^2$

$n = 800.33$ or 801

b. $\hat{p} = 0.5 \qquad \hat{q} = 0.5$

$n = \hat{p}\hat{q}\left[\frac{z_{\frac{\alpha}{2}}}{E}\right] = (0.5)(0.5)\left[\frac{1.96}{0.03}\right]^2$

$n = 1067.11$ or 1068

19.
$\hat{p} = 0.352 \qquad \hat{q} = 0.648$

$n = \hat{p}\hat{q}\left[\frac{z_{\frac{\alpha}{2}}}{E}\right]$

$n = (0.352)(0.648)\left[\frac{1.65}{0.025}\right]^2$

$n = 994$

21.
$600 = (0.5)(0.5)\left[\frac{z}{0.04}\right]^2$

$600 = 156.25 z^2$

$3.84 = z^2$

$\sqrt{3.84} = 1.96 = z$

1.96 corresponds to a 95% degree of confidence.

EXERCISE SET 7-4

1.
Chi-square (χ^2)

3.

	χ^2_{left}	χ^2_{right}
a.	3.816	21.920
b.	10.117	30.144
c.	13.844	41.923
d.	0.412	16.750
e.	26.509	55.758

Chapter 7 - Confidence Intervals and Sample Size

5.

$$\frac{(n-1)s^2}{\chi^2_{right}} < \sigma^2 < \frac{(n-1)s^2}{\chi^2_{left}}$$

$$\frac{16(10.1)^2}{28.845} < \sigma^2 < \frac{16(10.1)^2}{6.908}$$

$56.6 < \sigma^2 < 236.3$
$7.5 < \sigma < 15.4$

7.

$$\frac{(n-1)s^2}{\chi^2_{right}} < \sigma^2 < \frac{(n-1)s^2}{\chi^2_{left}}$$

$$\frac{19(1.6)^2}{32.852} < \sigma^2 < \frac{19(1.6)^2}{8.907}$$

$1.48 < \sigma^2 < 5.46$
$1.22 < \sigma < 2.34$

Yes, the estimate is reasonable.

9.

$s = 120.82$

$$\frac{(n-1)s^2}{\chi^2_{right}} < \sigma^2 < \frac{(n-1)s^2}{\chi^2_{left}}$$

$$\frac{14(120.82)^2}{23.685} < \sigma^2 < \frac{14(120.82)^2}{6.571}$$

$8{,}628.44 < \sigma^2 < 31{,}100.99$
$\$92.89 < \sigma < \$176.35$

11.

$$\frac{(n-1)s^2}{\chi^2_{right}} < \sigma^2 < \frac{(n-1)s^2}{\chi^2_{left}}$$

$$\frac{10(53)^2}{25.188} < \sigma^2 < \frac{10(53)^2}{2.156}$$

$1115.21 < \sigma^2 < 13{,}028.76$
$\$33.39 < \sigma < \$114.14$

13.

$s = 23.827$

$$\frac{(n-1)s^2}{\chi^2_{right}} < \sigma^2 < \frac{(n-1)s^2}{\chi^2_{left}}$$

$$\frac{13(23.827)^2}{24.736} < \sigma^2 < \frac{13(23.827)^2}{5.009}$$

$298.368 < \sigma^2 < 1473.435$
$17.3 < \sigma < 38.4$

15.

$s = 43.072$

$$\frac{(n-1)s^2}{\chi^2_{right}} < \sigma^2 < \frac{(n-1)s^2}{\chi^2_{left}}$$

$$\frac{14(43.072)^2}{26.119} < \sigma^2 < \frac{14(43.072)^2}{5.629}$$

$994.401 < \sigma^2 < 4614.099$
$31.5 < \sigma < 67.9$

REVIEW EXERCISES - CHAPTER 7

1.

$\overline{X} = 25$ is the point estimate of μ.

$$\overline{X} - z_{\frac{\alpha}{2}}\left(\frac{\sigma}{\sqrt{n}}\right) < \mu < \overline{X} + z_{\frac{\alpha}{2}}\left(\frac{\sigma}{\sqrt{n}}\right)$$

$$25 - 1.65\left(\frac{4}{\sqrt{49}}\right) < \mu < 25 + 1.65\left(\frac{4}{\sqrt{49}}\right)$$

$25 - 0.9429 < \mu < 25 + 0.9429$
$24.06 < \mu < 25.94$ or $24 < \mu < 26$
(TI answer: $24.06 < \mu < 25.94$)

3.

$$n = \left[\frac{z_{\frac{\alpha}{2}} \cdot \sigma}{E}\right]^2 = \left[\frac{1.65(4.8)}{2}\right]^2$$

$n = (3.96)^2 = 15.68$ or 16 female students

5.

$\overline{X} = 82.64$ is the point estimate of μ.

$$\overline{X} - t_{\frac{\alpha}{2}}\left(\frac{s}{\sqrt{n}}\right) < \mu < \overline{X} + t_{\frac{\alpha}{2}}\left(\frac{s}{\sqrt{n}}\right)$$

$$82.64 - 2.228\left(\frac{8.49}{\sqrt{11}}\right) < \mu < 82.64 + 2.228\left(\frac{8.49}{\sqrt{11}}\right)$$

$82.64 - 5.7 < \mu < 82.64 + 5.7$
$76.9 < \mu < 88.3$

7.

$$\hat{p} - (z_{\frac{\alpha}{2}})\sqrt{\frac{\hat{p}\hat{q}}{n}} < p < \hat{p} + (z_{\frac{\alpha}{2}})\sqrt{\frac{\hat{p}\hat{q}}{n}}$$

$$0.34 - 1.96\sqrt{\frac{(0.34)(0.66)}{1000}} < P <$$
$$0.34 + 1.96\sqrt{\frac{(0.34)(0.66)}{1000}}$$

$0.34 - 0.029 < p < 0.34 + 0.029$
$0.311 < p < 0.369$

Chapter 7 - Confidence Intervals and Sample Size

9.

$$\hat{p} = \frac{316}{600} = 0.5267 \qquad \hat{q} = \frac{284}{600} = 0.4733$$

$$\hat{p} - (z_{\frac{\alpha}{2}})\sqrt{\frac{\hat{p}\hat{q}}{n}} < p < \hat{p} + (z_{\frac{\alpha}{2}})\sqrt{\frac{\hat{p}\hat{q}}{n}}$$

$$0.5267 - 2.58\sqrt{\frac{(0.5267)(0.4733)}{600}} < p <$$
$$0.5267 + 2.58\sqrt{\frac{(0.5267)(0.4763)}{600}}$$

$0.5267 - 0.053 < p < 0.5267 + 0.053$

$0.474 < p < 0.580$

11.
$\hat{p} = 0.84 \qquad \hat{q} = 0.16$

$$n = \hat{p}\hat{q}\left[\frac{z_{\frac{\alpha}{2}}}{E}\right]^2 = (0.84)(0.16)\left[\frac{1.65}{0.03}\right]^2$$

$n = 406.56$ or 407

13.

$$\frac{(n-1)s^2}{\chi^2_{right}} < \sigma^2 < \frac{(n-1)s^2}{\chi^2_{left}}$$

$$\frac{(18-1)(0.29)^2}{30.191} < \sigma^2 < \frac{(18-1)(0.29)^2}{7.564}$$

$0.0474 < \sigma^2 < 0.1890$

$0.218 < \sigma < 0.435$ or $0.22 < \sigma < 0.44$

Yes; it seems that there is a large standard deviation.

15.

$$\frac{(n-1)s^2}{\chi^2_{right}} < \sigma^2 < \frac{(n-1)s^2}{\chi^2_{left}}$$

$$\frac{(15-1)(8.6)}{23.685} < \sigma^2 < \frac{(15-1)(8.6)}{6.571}$$

$5.1 < \sigma^2 < 18.3$

CHAPTER 7 QUIZ

1. True

3. False, it is consistent if, as sample size increases, the estimator approaches the parameter being estimated.

5. b

7. b

9. Margin of error

11. 90, 95, 99

13.
$\overline{X} = \$44.80$ is the point estimate for μ.

$$\overline{X} - t_{\frac{\alpha}{2}}\left(\frac{s}{\sqrt{n}}\right) < \mu < \overline{X} + t_{\frac{\alpha}{2}}\left(\frac{s}{\sqrt{n}}\right)$$

$$\$44.80 - 2.093\left(\frac{3.53}{\sqrt{20}}\right) < \mu < \$44.80 + 2.093\left(\frac{3.53}{\sqrt{20}}\right)$$

$\$43.15 < \mu < \$46.45$

15.

$$\overline{X} - t_{\frac{\alpha}{2}}\left(\frac{s}{\sqrt{n}}\right) < \mu < \overline{X} + t_{\frac{\alpha}{2}}\left(\frac{s}{\sqrt{n}}\right)$$

$$48.6 - 2.262\left(\frac{4.1}{\sqrt{10}}\right) < \mu < 48.6 + 2.262\left(\frac{4.1}{\sqrt{10}}\right)$$

$45.7 < \mu < 51.5$

17.

$$\overline{X} - t_{\frac{\alpha}{2}}\left(\frac{s}{\sqrt{n}}\right) < \mu < \overline{X} + t_{\frac{\alpha}{2}}\left(\frac{s}{\sqrt{n}}\right)$$

$$31 - 2.353\left(\frac{4}{\sqrt{4}}\right) < \mu < 31 + 2.353\left(\frac{4}{\sqrt{4}}\right)$$

$26 < \mu < 36$

19.

$$n = \left[\frac{z_{\frac{\alpha}{2}} \sigma}{E}\right]^2 = \left[\frac{1.65(900)}{300}\right]^2$$

$n = 24.5$ or 25

21.

$$p\hat{p} - (z_{\frac{\alpha}{2}})\sqrt{\frac{\hat{p}\hat{q}}{n}} < p < \hat{p} + (z_{\frac{\alpha}{2}})\sqrt{\frac{\hat{p}\hat{q}}{n}}$$

$$0.36 - 1.65\sqrt{\frac{(0.36)(0.64)}{150}} < p <$$
$$0.36 + 1.65\sqrt{\frac{(0.36)(0.64)}{150}}$$

$0.295 < p < 0.425$

23.

$$n = \hat{p}\hat{q}\left[\frac{z_{\frac{\alpha}{2}}}{E}\right]^2$$

$$n = (0.15)(0.85)\left[\frac{1.96}{0.03}\right]^2$$

n = 544.22 or 545

25.

$$\frac{(n-1)s^2}{\chi^2_{right}} < \sigma^2 < \frac{(n-1)s^2}{\chi^2_{left}}$$

$$\frac{26(6.8)^2}{38.885} < \sigma^2 < \frac{26(6.8)^2}{15.379}$$

$30.9 < \sigma^2 < 78.2$

$5.6 < \sigma < 8.8$

Chapter 8 - Hypothesis Testing

Note: Graphs are not to scale and are intended to convey a general idea. Answers may vary due to rounding.

EXERCISE SET 8-1

1.
The null hypothesis is a statistical hypothesis that states there is no difference between a parameter and a specific value or there is no difference between two parameters. The alternative hypothesis specifies a specific difference between a parameter and a specific value, or that there is a difference between two parameters. Examples will vary.

3.
A statistical test uses the data obtained from a sample to make a decision as to whether or not the null hypothesis should be rejected.

5.
The critical region is the region of values of the test-statistic that indicates a significant difference and the null hypothesis should be rejected. The non-critical region is the region of values of the test-statistic that indicates the difference was probably due to chance, and the null hypothesis should not be rejected.

7.
Type I is represented by α, type II is represented by β.

9.
A one-tailed test should be used when a specific direction, such as greater than or less than, is being hypothesized, whereas when no direction is specified, a two-tailed test should be used.

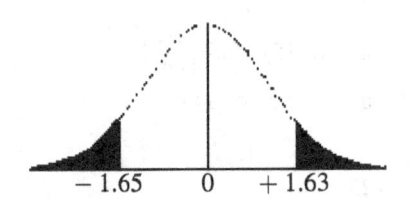

11.
a. ± 1.65

b. $+2.33$

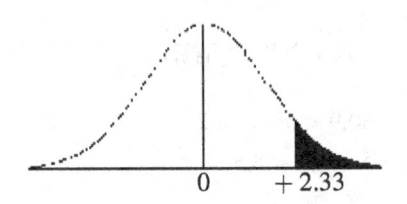

c. -2.58

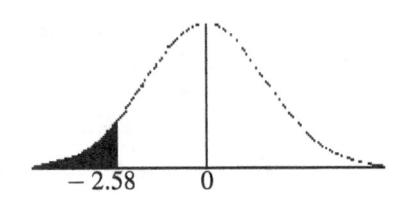

d. -2.33

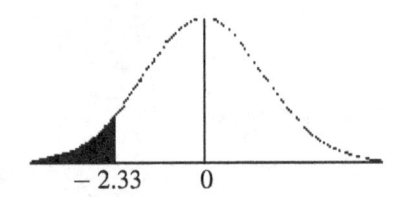

e. $+1.65$

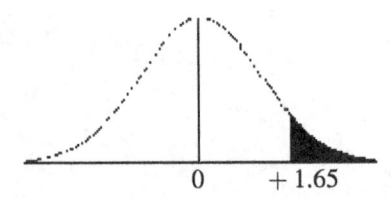

13.
a. $H_0: \mu = 15.6$
   $H_1: \mu \neq 15.6$

13. continued

b. $H_0: \mu = 10.8$
   $H_1: \mu > 10.8$

c. $H_0: \mu = 390$
   $H_1: \mu > 390$

d. $H_0: \mu = 12{,}603$
   $H_1: \mu \neq 12{,}603$

e. $H_0: \mu = 24$
   $H_1: \mu > 24$

EXERCISE SET 8-2

1.
$H_0: \mu = 305$

$H_1: \mu > 305$ (claim)

C. V. $= 1.65 \qquad \sigma = \sqrt{3.6} = 1.897$

$z = \dfrac{\overline{X} - \mu}{\frac{\sigma}{\sqrt{n}}} = \dfrac{306.2 - 305}{\frac{1.897}{\sqrt{55}}} = 4.69$

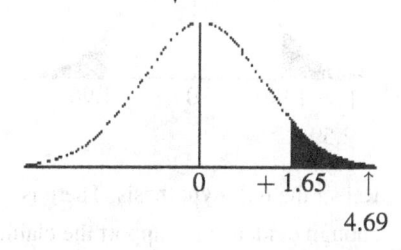

Reject the null hypothesis. There is enough evidence to support the claim that the mean depth is greater than 305 feet. Many factors could contribute to the increase, including warmer temperatures and higher than usual rainfall.

3.
$H_0: \mu = \$24$ billion
$H_1: \mu > \$24$ billion (claim)

C. V. $= +1.65 \quad \overline{X} = \$31.5 \quad \sigma = \$28.7$

$z = \dfrac{\overline{X} - \mu}{\frac{\sigma}{\sqrt{n}}} = \dfrac{31.5 - 24}{\frac{28.7}{\sqrt{50}}} = 1.85$

3. continued

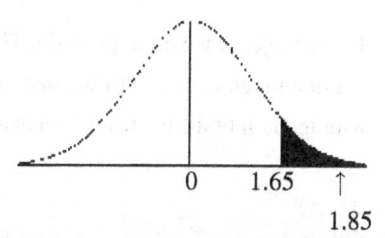

Reject the null hypothesis. There is enough evidence to support the claim that the average revenue exceeds $24 billion.

5.
$H_0: \mu = 5$
$H_1: \mu > 5$ (claim)

C. V. $= 2.33$

$z = \dfrac{\overline{X} - \mu}{\frac{\sigma}{\sqrt{n}}} = \dfrac{5.6 - 5}{\frac{1.2}{\sqrt{32}}} = 2.83$

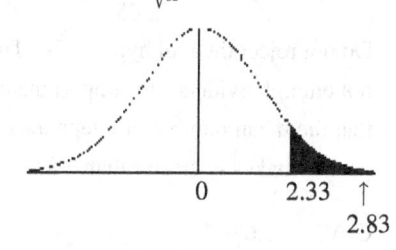

Reject the null hypothesis. There is enough evidence to support the claim that the mean number of sick days a person takes is greater than 5.

7.
$H_0: \mu = 29$
$H_1: \mu \neq 29$ (claim)

C. V. $= \pm 1.96 \quad \overline{X} = 29.45 \quad \sigma = 2.61$

$z = \dfrac{\overline{X} - \mu}{\frac{\sigma}{\sqrt{n}}} = \dfrac{29.45 - 29}{\frac{2.61}{\sqrt{30}}} = 0.944$

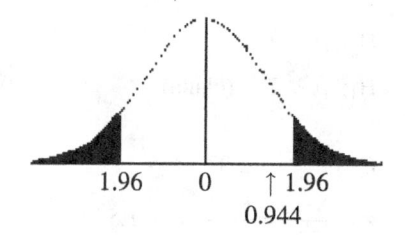

Chapter 8 - Hypothesis Testing

7. continued

Do not reject the null hypothesis. There is not enough evidence to say that the average height differs from 29 inches.

9.
$H_0: \mu = 2.8$
$H_1: \mu > 2.8$ (claim)

C. V. $= 2.33$
$z = \dfrac{\overline{X} - \mu}{\frac{\sigma}{\sqrt{n}}} = \dfrac{3.1 - 2.8}{\frac{0.8}{\sqrt{30}}} = 2.05$

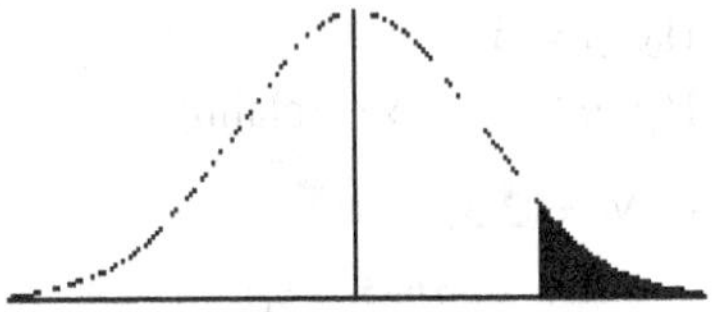

0 ↑ 2.33
2.05

Do not reject the null hypothesis. There is not enough evidence to support the claim that the mean number of telephone calls a person makes is greater than 2.8.

C. V. $= 1.65$

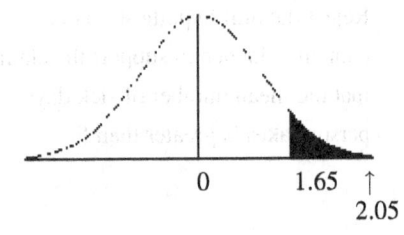

0   1.65 ↑
         2.05

Reject the null hypothesis. There is enough evidence to support the claim that the mean number of telephone calls a person makes is greater than 2.8.

11.
$H_0: \mu = 7$
$H_1: \mu < 7$ (claim)

C. V. $= -2.33$
$z = \dfrac{\overline{X} - \mu}{\frac{\sigma}{\sqrt{n}}} = \dfrac{6.5 - 7}{\frac{1.8}{\sqrt{32}}} = -1.57$

11. continued

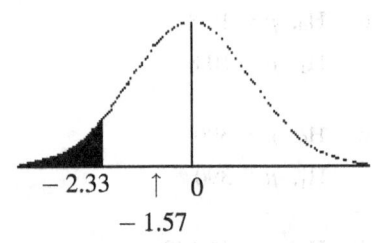

−2.33 ↑ 0
   −1.57

Do not reject the null hypothesis. There is not enough evidence to support the claim that the average weight loss of a newborn baby is less than 7 ounces in the first 2 days of life.

13.
$H_0: \mu = 15$
$H_1: \mu \neq 15$ (claim)
C. V. $= \pm 1.96$
$z = \dfrac{\overline{X} - \mu}{\frac{\sigma}{\sqrt{n}}} = \dfrac{13.8 - 15}{\frac{3}{\sqrt{42}}} = -2.59$

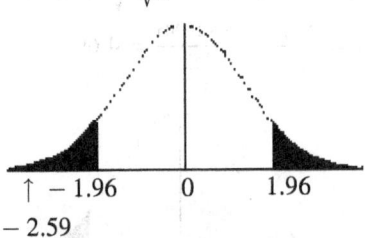

↑ −1.96   0   1.96
−2.59

Reject the null hypothesis. There is enough evidence to support the claim that the average differs from 15 shirts.

15.
a. Do not reject.
b. Reject.
c. Do not reject.
d. Reject.
e. Reject.

17.
$H_0: \mu = 264$
$H_1: \mu < 264$ (claim)
$z = \dfrac{\overline{X} - \mu}{\frac{\sigma}{\sqrt{n}}} = \dfrac{262.3 - 264}{\frac{3}{\sqrt{20}}} = -2.53$

17. continued

The area corresponding to z = 2.53 is 0.9943. The P-value is 1 − 0.9943 = 0.0057. The decision is to reject the null hypothesis since 0.0057 > 0.01. There is enough evidence to support the claim that the average stopping distance is less than 264 feet. (TI: P-value ≈ 0.0056)

19.

H_0: $\mu = 7.8$

H_1: $\mu > 7.8$ (claim)

$z = \dfrac{\bar{X} - \mu}{\frac{\sigma}{\sqrt{n}}} = \dfrac{8.7 - 7.8}{\frac{2.6}{\sqrt{35}}} = 2.05$

The area corresponding to z = 2.05 is 0.97985.
Thus, P-value = 1 − 0.9798 = 0.0202.
The decision is do not reject the null hypothesis since 0.0202 < 0.01. There is not enough evidence to support the claim that the average number of applications a potential medical school student is higher than 7.8. (TI: P-value = 0.0202)

21.

H_0: $\mu = 444$

H_1: $\mu \neq 444$

$z = \dfrac{\bar{X} - \mu}{\frac{\sigma}{\sqrt{n}}} = \dfrac{430 - 444}{\frac{52}{\sqrt{40}}} = -1.70$

The area corresponding to z = − 1.70 is 0.0446. The P-value is 2(0.0446) = 0.0892. The decision is do not reject the null hypothesis since P-value < 0.05. There is not enough evidence to support the claim that the mean differs from 444.
(TI: P-value = 0.0886)

23.

H_0: $\mu = 30,000$ (claim)

H_1: $\mu \neq 30,000$

$z = \dfrac{\bar{X} - \mu}{\frac{\sigma}{\sqrt{n}}} = \dfrac{30,456 - 30,000}{\frac{1684}{\sqrt{40}}} = 1.71$

23. continued

The area corresponding to z = 1.71 is 0.9564. The P-value is 2(1 − 0.9564) = 2(0.0436) = 0.0872. The decision is to reject the null hypothesis at $\alpha = 0.10$ since 0.0872 < 0.10. The conclusion is that there is enough evidence to reject the claim that customers are adhering to the recommendation. Yes, the 0.10 significance level is appropriate.
(TI: P-value = 0.0868)

25.

H_0: $\mu = 10$

H_1: $\mu < 10$ (claim)

$\bar{X} = 5.025$ $s = 3.63$

$z = \dfrac{\bar{X} - \mu}{\frac{\sigma}{\sqrt{n}}} = \dfrac{5.025 - 10}{\frac{3.63}{\sqrt{40}}} = -8.67$

The area corresponding to − 8.67 is less than 0.0001. Since 0.0001 < 0.05, the decision is to reject the null hypothesis. There is enough evidence to support the claim that the average number of days missed per year is less than 10.

27.

The mean and standard deviation are found as follows:

	f	X_m	$f \cdot X_m$	$f \cdot X_m^2$
8.35 - 8.43	2	8.39	16.78	140.7842
8.44 - 8.52	6	8.48	50.88	431.4624
8.53 - 8.61	12	8.57	102.84	881.3388
8.62 - 8.70	18	8.66	155.88	1349.9208
8.71 - 8.79	10	8.75	87.5	765.625
8.80 - 8.88	2	8.84	17.68	156.2912
	50		431.56	3725.4224

$\bar{X} = \dfrac{\sum f \cdot X_m}{n} = \dfrac{431.56}{50} = 8.63$

27. continued

$$s = \sqrt{\frac{\sum f \cdot X_m^2 - \frac{(\sum f \cdot X_m)^2}{n}}{n-1}} = \sqrt{\frac{3725.4224 - \frac{(431.56)^2}{50}}{49}}$$

$s = 0.105$

H_0: $\mu = 8.65$ (claim)

H_1: $\mu \neq 8.65$

C. V. $= \pm 1.96$

$z = \dfrac{\overline{X} - \mu}{\frac{s}{\sqrt{n}}} = \dfrac{8.63 - 8.65}{\frac{0.105}{\sqrt{50}}} = -1.35$

Do not reject the null hypothesis. There is not enough evidence to reject the claim that the average hourly wage of the employees is $8.65.

EXERCISE SET 8-3

1.
It is bell-shaped, symmetric about the mean, and it never touches the x axis. The mean, median, and mode are all equal to 0 and they are located at the center of the distribution. The t distribution differs from the standard normal distribution in that it is a family of curves, the variance is greater than 1, and as the degrees of freedom increase the t distribution approaches the standard normal distribution.

3.
a. d. f. = 11   C. V. = -2.718
b. d. f. = 15   C. V. = $+1.753$
c. d. f. = 6   C. V. = ± 1.943
d. d. f. = 10   C. V. = $+2.228$
e. d. f. = 9   C. V. = ± 2.262

5.
a. $0.01 < $ P-value < 0.025 (0.018)
b. $0.05 < $ P-value < 0.10 (0.062)
c. $0.10 < $ P-value < 0.25 (0.123)
d. $0.10 < $ P-value < 0.20 (0.138)

7.
H_0: $\mu = 31$

H_1: $\mu < 31$ (claim)

C. V. $= -1.833$ d. f. $= 9$

$t = \dfrac{\overline{X} - \mu}{\frac{s}{\sqrt{n}}} = \dfrac{28 - 31}{\frac{2.7}{\sqrt{10}}} = -3.514$

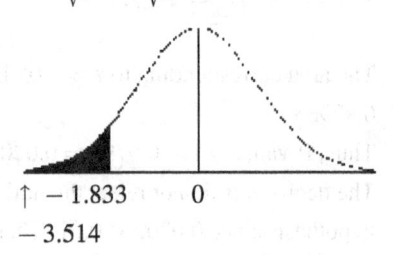

↑ -1.833 0
-3.514

Reject the null hypothesis. There is enough evidence to support the claim that the mean number of cigarettes that smokers smoke is less than 31 per day.

9.
H_0: $\mu = 700$ (claim)

H_1: $\mu < 700$

$\overline{X} = 606.5$ $s = 109.1$

C. V. $= -2.262$ d. f. $= 9$

$t = \dfrac{\overline{X} - \mu}{\frac{s}{\sqrt{n}}} = \dfrac{606.5 - 700}{\frac{109.1}{\sqrt{10}}} = -2.710$

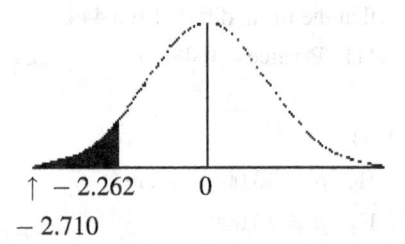

↑ -2.262 0
-2.710

Chapter 8 - Hypothesis Testing

9. continued

Reject the null hypothesis. There is enough evidence to reject the claim that the average height of the buildings is at least 700 feet.

11.
$H_0: \mu = 58$
$H_1: \mu > 58$ (claim)

C. V. $= 2.821$ d. f. $= 9$
$\overline{X} = 123.5$ $s = 39.303$
$t = \dfrac{\overline{X} - \mu}{\frac{s}{\sqrt{n}}} = \dfrac{123.5 - 58}{\frac{39.303}{\sqrt{10}}} = 5.27$

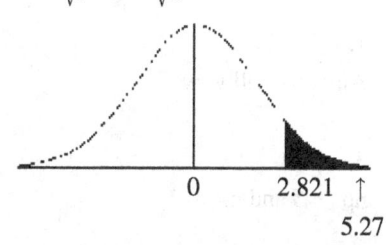

0   2.821 ↑
         5.27

Reject the null hypothesis. There is enough evidence to support the claim that the average is greater than the national average.

13.
$H_0: \mu = 7.2$
$H_1: \mu \neq 7.2$ (claim)
C. V. $= \pm 2.145$ d. f. $= 14$
$t = \dfrac{\overline{X} - \mu}{\frac{s}{\sqrt{n}}} = \dfrac{8.3 - 7.2}{\frac{1.2}{\sqrt{15}}} = 3.550$

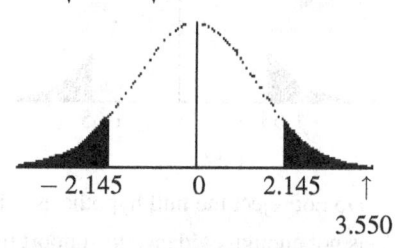

−2.145   0   2.145 ↑
                    3.550

Reject the null hypothesis. There is enough evidence to support the claim that the mean number of hours that college students sleep on Friday night to Saturday morning is not 7.2 hours.

15.
$H_0: \mu = \$50.07$
$H_1: \mu > \$50.07$ (claim)

C. V. $= 1.833$ d. f. $= 9$
$\overline{X} = \$56.11$ $s = \$6.97$
$t = \dfrac{\overline{X} - \mu}{\frac{s}{\sqrt{n}}} = \dfrac{\$56.11 - \$50.07}{\frac{6.97}{\sqrt{10}}} = 2.741$

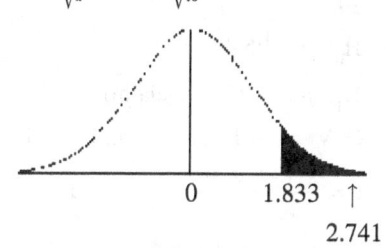

0   1.833 ↑
         2.741

Reject the null hypothesis. There is enough evidence to support the claim that the average bill has increased.

17.
$H_0: \mu = 211$
$H_1: \mu > 211$ (claim)
C. V. $= -1.345$ d. f. $= 14$
$t = \dfrac{\overline{X} - \mu}{\frac{s}{\sqrt{n}}} = \dfrac{208.8 - 211}{\frac{3.8}{\sqrt{15}}} = -2.242$

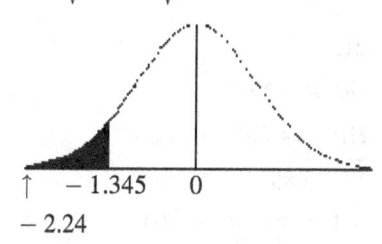

↑ −1.345   0
−2.24

Reject the null hypothesis. There is enough evidence to support director's feelings that his surgeons perform fewer operations per year than the national average of 211.

C. V. $= -2.624$ d. f. $= 14$

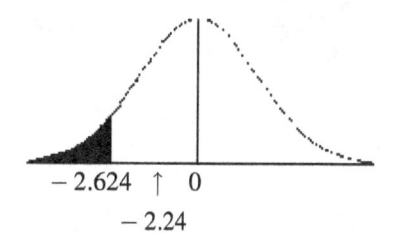

−2.624 ↑ 0
−2.24

17. continued

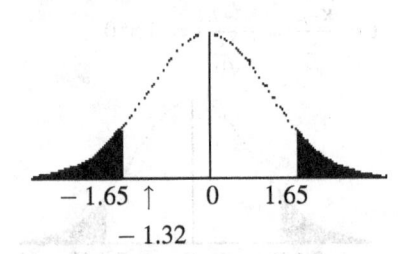

Do not reject the null hypothesis. There is not enough evidence to support director's feelings that his surgeons perform fewer operations per year than the national average of 211.

19.

$H_0: \mu = 25.4$
$H_1: \mu > 25.4$ (claim)
C. V. $= -1.318$ d. f. $= 24$
$t = \dfrac{\overline{X} - \mu}{\frac{s}{\sqrt{n}}} = \dfrac{22.1 - 25.4}{\frac{5.3}{\sqrt{25}}} = -3.11$

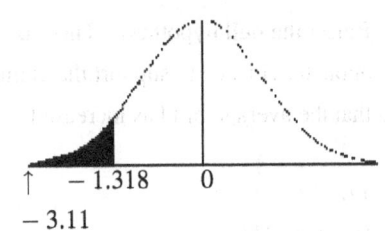

↑ -1.318 0
-3.11

Reject the null hypothesis. There is enough evidence to support the claim that the commute time is less than 25.4 minutes.

21.

$H_0: \mu = 5.8$
$H_1: \mu \neq 5.8$ (claim)
$\overline{X} = 3.85$ $s = 2.519$
d. f. $= 19$ $\alpha = 0.05$
P-value < 0.01 (0.0026)
$t = \dfrac{\overline{X} - \mu}{\frac{s}{\sqrt{n}}} = \dfrac{3.85 - 5.8}{\frac{2.519}{\sqrt{20}}} = -3.462$

Since P-value < 0.01, reject the null hypothesis. There is enough evidence to support the claim that the mean is not 5.8.

23.

$H_0: \mu = 123$
$H_1: \mu \neq 123$ (claim)
d. f. $= 15$
P-value < 0.01 (0.0086)
$t = \dfrac{\overline{X} - \mu}{\frac{s}{\sqrt{n}}} = \dfrac{119 - 123}{\frac{5.3}{\sqrt{16}}} = -3.019$

Since P-value < 0.05, reject the null hypothesis. There is enough evidence to support the claim that the mean is not 123 gallons.

EXERCISE SET 8-4

1.
Answers will vary.

3.
$np \geq 5$ and $nq \geq 5$

5.
$H_0: p = 0.46$
$H_1: p \neq 0.46$ (claim)
$\hat{p} = \dfrac{48}{120} = 0.4$ $p = 0.46$ $q = 0.54$
C. V. $= \pm 1.65$
$z = \dfrac{\hat{p} - p}{\sqrt{\frac{pq}{n}}} = \dfrac{0.4 - 0.46}{\sqrt{\frac{(0.46)(0.54)}{120}}} = -1.32$
(TI: $z = -1.32$)

-1.65 ↑   0 1.65
-1.32

Do not reject the null hypothesis. There is not enough evidence to support the claim that the percentage has changed.

7.

H_0: $p = 0.11$

H_1: $p \neq 0.11$ (claim)

$\hat{p} = \frac{32}{200} = 0.16$ $p = 0.11$ $q = 0.89$

C. V. $= \pm 2.33$

$z = \frac{\hat{p} - p}{\sqrt{\frac{pq}{n}}} = \frac{0.16 - 0.11}{\sqrt{\frac{(0.11)(0.89)}{200}}} = 2.26$

(TI: $z = 2.26$)

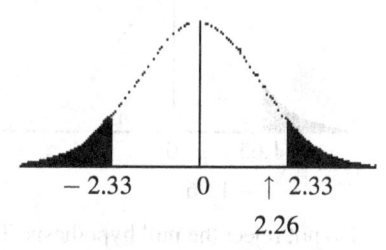

$-2.33 \quad 0 \quad \uparrow 2.33$
$\quad\quad\quad\quad\quad 2.26$

Do not reject the null hypothesis. There is not enough evidence to reject the claim that 11% of individuals eat takeout food every day.

C. V. $= \pm 1.96$

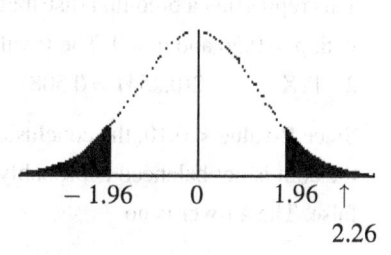

$-1.96 \quad 0 \quad 1.96 \uparrow$
$\quad\quad\quad\quad\quad 2.26$

Reject the null hypothesis. There is enough evidence to reject the claim that 11% of individuals eat takeout food every day.

9.

H_0: $p = 0.58$

H_1: $p > 0.58$ (claim)

$\hat{p} = \frac{63}{90} = 0.70$ $p = 0.58$ $q = 0.42$

C. V. $= 1.65$

$z = \frac{\hat{p} - p}{\sqrt{\frac{pq}{n}}} = \frac{0.70 - 0.58}{\sqrt{\frac{(0.58)(0.42)}{90}}} = 2.31$

(TI: $z = 2.31$)

9. continued

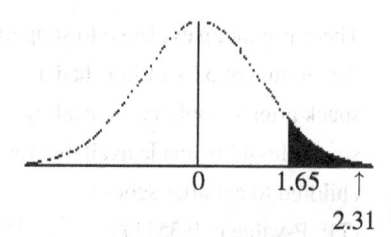

$0 \quad 1.65 \uparrow$
$\quad\quad\quad 2.31$

Reject the null hypothesis. There is enough evidence to conclude that the proportion of female runaways is higher than 58%.

11.

H_0: $p = 0.76$

H_1: $p < 0.76$ (claim)

$\hat{p} = \frac{38}{56} = 0.6786$ $p = 0.76$ $q = 0.24$

C. V. $= -2.33$

$z = \frac{\hat{p} - p}{\sqrt{\frac{pq}{n}}} = \frac{0.6786 - 0.76}{\sqrt{\frac{(0.76)(0.24)}{56}}} = -1.43$

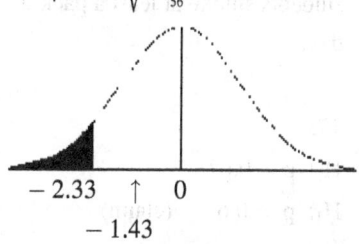

$-2.33 \quad \uparrow \quad 0$
$\quad\quad -1.43$

Do not reject the null hypothesis. There is not enough evidence to support the claim that the percentage is less than 76%.

13.

H_0: $p = 0.54$ (claim)

H_1: $p \neq 0.54$

$\hat{p} = \frac{36}{60} = 0.6$ $p = 0.54$ $q = 0.46$

$z = \frac{\hat{p} - p}{\sqrt{\frac{pq}{n}}} = \frac{0.6 - 0.54}{\sqrt{\frac{(0.54)(0.46)}{60}}} = 0.93$

Area $= 0.8238$

P-value $= 2(1 - 0.8238) = 0.3524$

Since P-value > 0.01, do not reject the null hypothesis.

13. continued

There is enough evidence to support the claim that 54% of kids had a snack after school. Yes, a healthy snack should be made available for children to eat after school.
(TI: P-value ≈ 0.3511)

15.

H_0: $p = 0.18$ (claim)
H_1: $p < 0.18$

$\hat{p} = \frac{50}{300} = 0.1667$ $p = 0.18$ $q = 0.82$

$z = \frac{\hat{p}-p}{\sqrt{\frac{pq}{n}}} = \frac{0.1667-0.18}{\sqrt{\frac{(0.18)(0.82)}{300}}} = -0.60$

P-value = 0.2743 (TI: P-value = 0.2739)
Since P-value > 0.05, do not reject the null hypothesis. There is not enough evidence to reject the claim that 18% of all high school students smoke at least a pack of cigarettes a day.

17.

H_0: $p = 0.67$
H_1: $p \neq 0.67$ (claim)

$\hat{p} = \frac{82}{100} = 0.82$ $p = 0.67$ $q = 0.33$

C. V. = ± 1.96

$z = \frac{\hat{p}-p}{\sqrt{\frac{pq}{n}}} = \frac{0.82-0.67}{\sqrt{\frac{(0.67)(0.33)}{100}}} = 3.19$

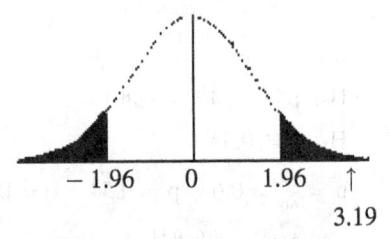

Reject the null hypothesis. There is enough evidence to support the claim that the percentage is not 67%.

19.

H_0: $p = 0.576$
H_1: $p < 0.576$ (claim)

$\hat{p} = \frac{17}{36} = 0.472$ $p = 0.576$ $q = 0.424$

C. V. = −1.65

$z = \frac{\hat{p}-p}{\sqrt{\frac{pq}{n}}} = \frac{0.472-0.576}{\sqrt{\frac{(0.576)(0.424)}{36}}} = -1.26$

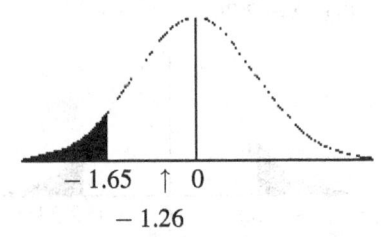

Do not reject the null hypothesis. There is not enough evidence to support the claim that the percentage of injuries during practice is less than 57.6%.

21.

This represents a binomial distribution with $p = 0.50$ and $n = 9$. The P-value is $2 \cdot P(X \leq 3) = 2(0.254) = 0.508$.

Since P-value < 0.10, the conclusion that the coin is not balanced is probably false. The answer is no.

23.

$z = \frac{X-\mu}{\sigma}$

$z = \frac{X-np}{\sqrt{npq}}$

$z = \frac{\frac{X}{n}-\frac{np}{n}}{\frac{1}{n}\sqrt{npq}}$

$z = \frac{\frac{X}{n}-\frac{np}{n}}{\sqrt{\frac{npq}{n^2}}}$

$z = \frac{\hat{p}-p}{\sqrt{\frac{pq}{n}}}$

Chapter 8 - Hypothesis Testing

EXERCISE SET 8-5

1.

a. $H_0: \sigma^2 = 225$
   $H_1: \sigma^2 \neq 225$
   C. V. = 5.892, 22.362   d. f. = 13

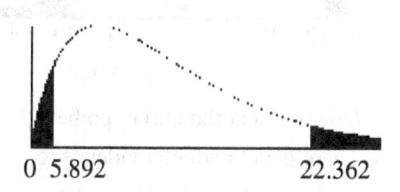

b. $H_0: \sigma^2 = 225$
   $H_1: \sigma^2 > 225$
   C. V. = 38.885   d. f. = 26

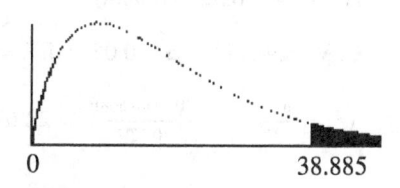

c. $H_0: \sigma^2 = 225$
   $H_1: \sigma^2 < 225$
   C. V. = 1.646   d. f. = 8

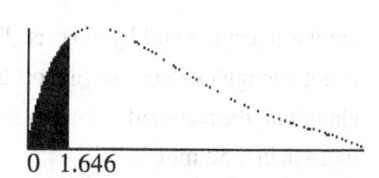

d. $H_0: \sigma^2 = 225$
   $H_1: \sigma^2 > 225$
   C. V. = 26.296   d. f. = 16

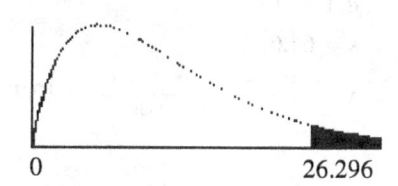

3.

a. 0.01 < P-value < 0.025 (0.015)
b. 0.005 < P-value < 0.01 (0.006)
c. 0.01 < P-value < 0.02 (0.012)
d. P-value < 0.005   (0.003)

5.

$H_0: \sigma = 8.6$ (claim)
$H_1: \sigma \neq 8.6$

$s^2 = 86.49$

C. V. = 3.816, 21.920 $\alpha = 0.05$
d. f. = 11

$$\chi^2 = \frac{(n-1)s^2}{\sigma^2} = \frac{(12-1)(86.49)}{73.96} = 12.864$$

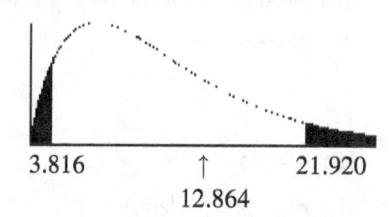

Do not reject the null hypothesis. There is not enough evidence to reject the claim that the standard deviation of the ages is 8.6 years.

7.

$H_0: \sigma = 1.2$ (claim)
$H_1: \sigma > 1.2$

$\alpha = 0.01$ d. f. = 14

$$\chi^2 = \frac{(n-1)s^2}{\sigma^2} = \frac{(15-1)(1.8)^2}{(1.2)^2} = 31.5$$

P-value < 0.005   (0.0047)

Since P-value < 0.01, reject the null hypothesis. There is enough evidence to reject the claim that the standard deviation is less than or equal to 1.2 minutes.

9.

H_0: $\sigma = 2$

H_1: $\sigma > 2$ (claim)

$s = 2.6375$

C. V. = 12.017 $\alpha = 0.10$ d. f. = 7

$\chi^2 = \frac{(n-1)s^2}{\sigma^2} = \frac{(8-1)(2.6375)^2}{2^2} = 12.174$

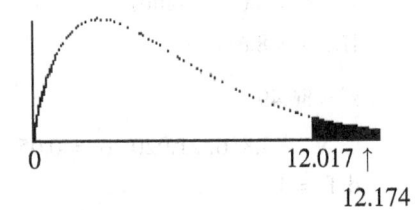

0   12.017 ↑
                12.174

Reject the null hypothesis. There is enough evidence to support the claim that the standard deviation is greater than two miles.

11.

H_0: $\sigma = 35$

H_1: $\sigma > 35$ (claim)

C. V. = 3.940 $\alpha = 0.05$ d. f. = 10

$\chi^2 = \frac{(n-1)s^2}{\sigma^2} = \frac{(11-1)(32)^2}{35^2} = 8.359$

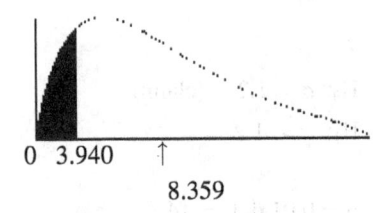

0 3.940 ↑
          8.359

Do not reject the null hypothesis. There is not enough evidence to support the claim that the standard deviation is less than 35.

13.

H_0: $\sigma^2 = 0.638$ (claim)

H_1: $\sigma^2 \neq 0.638$

C. V. = 12.401, 39.364 $\alpha = 0.05$

d. f. = 24

$s^2 = 0.930$

13. continued

$\chi^2 = \frac{(n-1)s^2}{\sigma^2} = \frac{(25-1)(0.930)}{0.638} = 34.984$

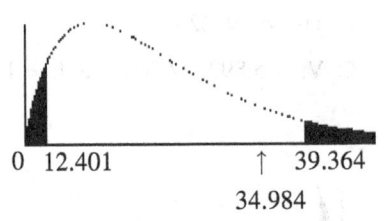

0 12.401   ↑ 39.364
                    34.984

Do not reject the null hypothesis. There is not enough evidence to reject the claim that the variance is equal to 0.638.

15.

H_0: $\sigma = 0.52$

H_1: $\sigma > 0.52$ (claim)

C. V. = 30.144 $\alpha = 0.05$ d. f. = 19

$\chi^2 = \frac{(n-1)s^2}{\sigma^2} = \frac{(20-1)(0.568)^2}{(0.52)^2} = 22.670$

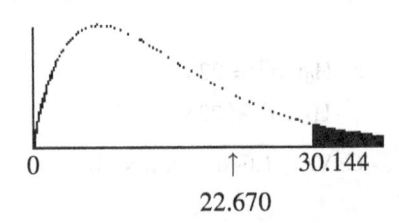

0   ↑   30.144
             22.670

Do not reject the null hypothesis. There is not enough evidence to support the claim that the standard deviation is more than 0.52 mm.

17.

H_0: $\sigma = 60$ (claim)

H_1: $\sigma \neq 60$

C. V. = 8.672, 27.587 $\alpha = 0.10$

d. f. = 17

$s = 64.6$

$\chi^2 = \frac{(n-1)s^2}{\sigma^2} = \frac{(18-1)(64.6)^2}{(60)^2} = 19.707$

17. continued

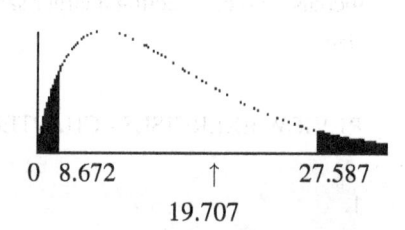

0   8.672   ↑   27.587
              19.707

Do not reject the null hypothesis. There is not enough evidence to reject the claim that the standard deviation is 60.

19.

$\sigma \approx \frac{Range}{4}$

$\sigma \approx \frac{\$9500-\$6782}{4} = \$679.50$

H_0: $\sigma = \$679.50$
H_1: $\sigma \neq \$679.50$ (claim)

s = 770.67

C. V. = 5.009, 24.736 $\alpha = 0.05$ d. f. = 13

$\chi^2 = \frac{(n-1)s^2}{\sigma^2} = \frac{(14-1)(770.67)^2}{679.5^2} = 16.723$

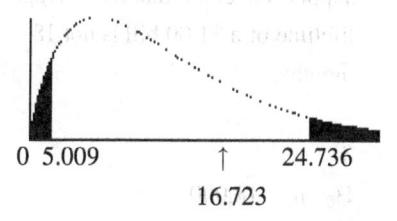

0   5.009   ↑   24.736
              16.723

Do not reject the null hypothesis. There is not enough evidence to support the claim that the standard deviation differs from $679.50.

EXERCISE SET 8-6

1.

H_0: $\mu = 25.2$
H_1: $\mu \neq 25.2$ (claim)

C. V. = ± 2.032

1. continued

$t = \frac{\overline{X}-\mu}{\frac{s}{\sqrt{n}}} = \frac{28.7-25.2}{\frac{4.6}{\sqrt{35}}} = 4.50$

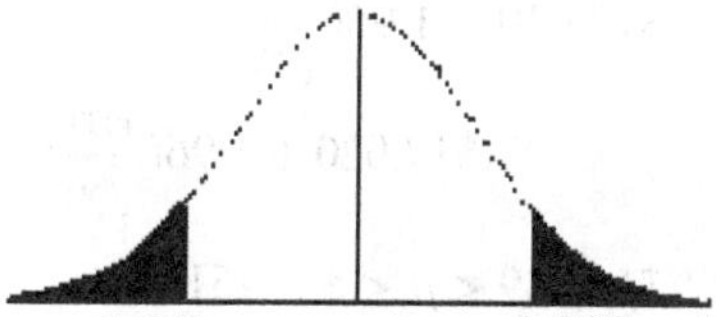

− 2.032   0   2.032   ↑
                              4.50

Reject the null hypothesis. There is enough evidence to support the claim that the average age differs from 25.2.

The 95% confidence interval of the mean is:

$\overline{X} - t_{\frac{\alpha}{2}}\frac{s}{\sqrt{n}} < \mu < \overline{X} + t_{\frac{\alpha}{2}}\frac{s}{\sqrt{n}}$

$28.7 - 2.032\left(\frac{4.6}{\sqrt{35}}\right) < \mu < 28.7 + 2.032\left(\frac{4.6}{\sqrt{35}}\right)$

$27.1 < \mu < 30.3$

(TI: $27.2 < \mu < 30.2$)

The confidence interval does not contain the hypothesized mean age of 25.2.

3.

H_0: $\mu = \$19,150$
H_1: $\mu \neq \$19,150$ (claim)

C. V. = ± 1.96

$z = \frac{\overline{X}-\mu}{\frac{\sigma}{\sqrt{n}}} = \frac{\$17,020-\$19,150}{\frac{4080}{\sqrt{50}}} = -3.69$

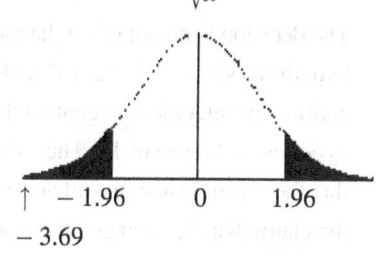

↑   − 1.96   0   1.96
− 3.69

Reject the null hypothesis. There is enough evidence to support the claim that the mean is not $19,150.

3. continued

$$\overline{X} - z_{\frac{\alpha}{2}}\frac{\sigma}{\sqrt{n}} < \mu < \overline{X} + z_{\frac{\alpha}{2}}\frac{\sigma}{\sqrt{n}}$$

$$\$17{,}020 - 1.96\left(\frac{4080}{\sqrt{50}}\right) < \mu <$$
$$\$17{,}020 + 1.96\left(\frac{4080}{\sqrt{50}}\right)$$

$\$15{,}889 < \mu < \$18{,}151$

The 95% confidence interval supports the conclusion because it does not contain the hypothesized mean.

5.

H_0: $\mu = 19$
H_1: $\mu \neq 19$ (claim)

C. V. $= \pm 2.145$

$$t = \frac{\overline{X}-\mu}{\frac{s}{\sqrt{n}}} = \frac{21.3-19}{\frac{6.5}{\sqrt{15}}} = 1.37$$

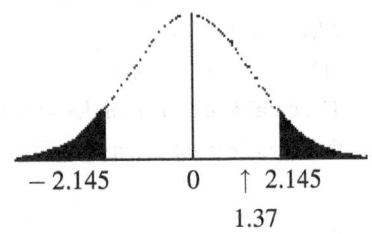

−2.145   0   ↑ 2.145
              1.37

The 99% confidence interval of the mean is:

$$\overline{X} - z_{\frac{\alpha}{2}}\frac{\sigma}{\sqrt{n}} < \mu < \overline{X} + z_{\frac{\alpha}{2}}\frac{\sigma}{\sqrt{n}}$$

$$21.3 - 2.145 \cdot \frac{6.5}{\sqrt{15}} < \mu < 21.3 + 2.145 \cdot \frac{6.5}{\sqrt{15}}$$

$17.7 < \mu < 24.9$

The decision is do not reject the null hypothesis since $1.37 < 2.145$ and the 99% confidence interval does contain the hypothesized mean of 19. The conclusion is that there is not enough evidence to support the claim that the average time worked at home is not 19 hours per week.

7.

The power of a statistical test is the probability of rejecting the null hypothesis when it is false.

9.

The power of a test can be increased by increasing α or selecting a larger sample size.

REVIEW EXERCISES - CHAPTER 8

1.

H_0: $\mu = 18$
H_1: $\mu \neq 18$ (claim)

C. V. $= \pm 2.33$
$\sigma = 2.8$

$$z = \frac{\overline{X}-\mu}{\frac{\sigma}{\sqrt{n}}} = \frac{18.8-18}{\frac{2.8}{\sqrt{50}}} = 2.02$$

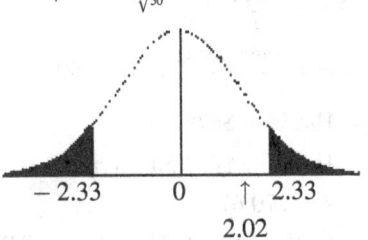

−2.33   0   ↑ 2.33
                2.02

Do not reject the null hypothesis. There is not enough evidence to support the claim that the average lifetime of a $1.00 bill is not 18 months.

3.

H_0: $\mu = 18{,}000$
H_1: $\mu < 18{,}000$ (claim)

$\overline{X} = 16{,}298.37$ $s = 2604.82$

C. V. $= -2.33$

$$z = \frac{\overline{X}-\mu}{\frac{s}{\sqrt{n}}} = \frac{16{,}298.37-18{,}000}{\frac{2604.82}{\sqrt{30}}} = -3.58$$

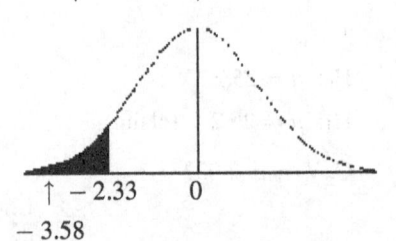

↑ −2.33   0
−3.58

Chapter 8 - Hypothesis Testing

3. continued

Reject the null hypothesis. There is enough evidence to support the claim that average debt is less than $18,000.

5.
H_0: $\mu = 22$
H_1: $\mu < 22$ (claim)

C. V. = 1.65
$\overline{X} = 23.2$ $\sigma = 3.7$

$z = \dfrac{\overline{X} - \mu}{\frac{\sigma}{\sqrt{n}}} = \dfrac{23.2 - 22}{\frac{3.7}{\sqrt{36}}} = 1.95$

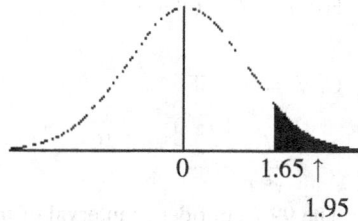

Reject the null hypothesis. There is enough evidence to support the claim that the mean is greater than 22 items.

7.
H_0: $\mu = 10$
H_1: $\mu > 10$ (claim)

C. V. = -1.782 $\overline{X} = 9.6385$ $s = 0.5853$

$t = \dfrac{\overline{X} - \mu}{\frac{s}{\sqrt{n}}} = \dfrac{9.6385 - 10}{\frac{0.5853}{\sqrt{13}}} = -2.227$ or -2.230

Reject the null hypothesis. There is enough evidence to support the claim that average weight is less than 10 ounces.

9.
H_0: $p = 0.25$
H_1: $p > 0.25$ (claim)

C. V. = -1.65

$\hat{p} = 0.19$ $p = 0.25$ $q = 0.75$

$z = \dfrac{\hat{p} - p}{\sqrt{\frac{pq}{n}}} = \dfrac{0.19 - 0.25}{\sqrt{\frac{(0.25)(0.75)}{100}}} = -1.39$

Do not reject the null hypothesis. There is not enough evidence to support the claim that less than 25% medical doctors received their degrees from foreign schools.

11.
H_0: $p = 0.593$
H_1: $p < 0.593$ (claim)

C. V. = -2.33

$\hat{p} = \dfrac{156}{300} = 0.52$ $p = 0.593$ $q = 0.407$

$z = \dfrac{\hat{p} - p}{\sqrt{\frac{pq}{n}}} = \dfrac{0.52 - 0.593}{\sqrt{\frac{(0.593)(0.407)}{300}}} = -2.57$

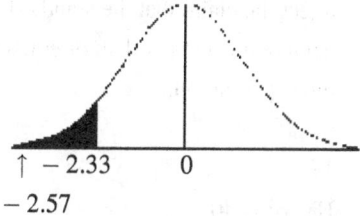

Reject the null hypothesis. There is enough evidence to support the claim that less than 59.3% of school lunches are free or at a reduced price.

Chapter 8 - Hypothesis Testing

13.
H_0: $p = 0.204$
H_1: $p \neq 0.204$ (claim)
$\hat{p} = 0.18$ $p = 0.204$ $q = 0.796$
C. V. $= \pm 1.96$

$$z = \frac{\hat{p}-p}{\sqrt{\frac{pq}{n}}} = \frac{0.18-0.204}{\sqrt{\frac{(0.204)(0.796)}{300}}} = -1.03$$

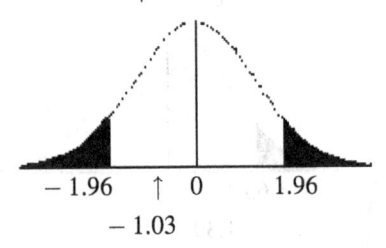

−1.96 ↑ 0   1.96
    −1.03

Do not reject the null hypothesis. There is not enough evidence to support the claim that the proportion of high school smokers differs from 20.4%.

15.
H_0: $\sigma = 4.3$ (claim)
H_1: $\sigma < 4.3$
d. f. $= 19$

$$\chi^2 = \frac{(n-1)s^2}{\sigma^2} = \frac{(20-1)(2.6)^2}{(4.3)^2} = 6.95$$

$0.005 <$ P-value < 0.01 (0.006) Since P-value < 0.05, reject the null hypothesis. There is enough evidence to reject the claim that the standard deviation is greater than or equal to 4.3 miles per gallon.

17.
H_0: $\sigma^2 = 40$
H_1: $\sigma^2 \neq 40$ (claim)
$s = 6.6$ $s^2 = (6.6)^2 = 43.56$ C.
V. $= 2.700, 19.023$ d. f. $= 9$

$$\chi^2 = \frac{(n-1)s^2}{\sigma^2} = \frac{(10-1)(43.56)}{40} = 9.801$$

17. continued

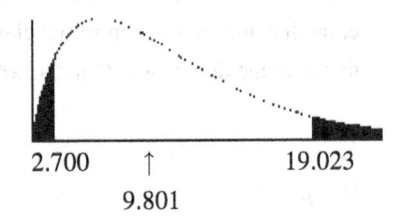

2.700 ↑   19.023
   9.801

Do not reject the null hypothesis. There is not enough evidence to support the claim that the variance differs from 40.

19.
H_0: $\mu = 4$
H_1: $\mu \neq 4$ (claim)

C. V. $= \pm 2.58$

$$z = \frac{\overline{X}-\mu}{\frac{s}{\sqrt{n}}} = \frac{4.2-4}{\frac{0.6}{\sqrt{20}}} = 1.49$$

The 99% confidence interval of the mean is:

$$\overline{X} - z_{\frac{\alpha}{2}}\frac{\sigma}{\sqrt{n}} < \mu < \overline{X} + z_{\frac{\alpha}{2}}\frac{\sigma}{\sqrt{n}}$$

$$4.2 - 2.58 \cdot \frac{0.6}{\sqrt{20}} < \mu < 4.2 + 2.58 \cdot \frac{0.6}{\sqrt{20}}$$

$3.85 < \mu < 4.55$

The decision is do not reject the null hypothesis since $1.49 < 2.58$ and the confidence interval does contain the hypothesized mean of 4. There is not enough evidence to support the claim that the growth has changed. Yes, the results agree. The hypothesized mean is contained in the interval.

CHAPTER 8 QUIZ

1. True

3. False, the critical value separates the critical region from the noncritical region.

5. False, it can be one-tailed or two-tailed.

Chapter 8 - Hypothesis Testing

7. d

9. b

11. β

13. Right

15. H_0: $\mu = 28.6$ (claim)
H_1: $\mu \neq 28.6$
C. V. $= \pm 1.96$
$z = 2.15$

Reject the null hypothesis. There is enough evidence to reject the claim that the average age is 28.6.

17. H_0: $\mu = 8$
H_1: $\mu > 8$ (claim)
C. V. $= 1.65$
$z = 6.00$
Reject the null hypothesis. There is enough evidence to support the claim that the average is greater than 8.

19. H_0: $\mu = 67$
H_1: $\mu > 67$ (claim)
$t = -3.1568$
P-value < 0.005 (0.003)
Since P-value < 0.05, reject the null hypothesis. There is enough evidence to support the claim that the average height is less than 67 inches.

21. H_0: $\mu = 63.5$
H_1: $\mu < 63.5$ (claim)
$t = 0.47075$
P-value > 0.25 (0.322)
Since P-value > 0.05, do not reject the null hypothesis. There is not enough evidence to support the claim that the average is greater than 63.5.

23. H_0: $p = 0.39$ (claim)
H_1: $p \neq 0.39$
C. V. $= \pm 1.96$
$z = -0.62$
Do not reject the null hypothesis. There is not enough evidence to reject the claim that 39% took supplements. The study supports the results of the previous study.

25. H_0: $p = 0.35$ (claim)
H_1: $p \neq 0.35$
C. V. $= \pm 2.33$
$z = 0.666$
Do not reject the null hypothesis. There is not enough evidence to reject the claim that the proportion is 35%.

27. The area corresponding to $z = 2.15$ is 0.9842.
P-value $= 2(1 - 0.9842) = 0.0316$

29. H_0: $\sigma = 6$
H_1: $\sigma < 6$ (claim)
C. V. $= 36.415$
$\chi^2 = 54$
Reject the null hypothesis. There is enough evidence to support the claim that the standard deviation is more than 6.

31. H_0: $\sigma = 2.3$
H_1: $\sigma < 2.3$ (claim)
C. V. $= 10.117$
$\chi^2 = 13$
Do not reject the null hypothesis. There is not enough evidence to support the claim that the standard deviation is less than 2.3.

33. $28.9 < \mu < 31.2$; no

Chapter 9 - Testing the Difference Between Two Means, Two Proportions, and Two Variances

Note: Graphs are not to scale and are intended to convey a general idea. Answers may vary due to rounding, TI-83's, or computer programs.

EXERCISE SET 9-1

1.
Testing a single mean involves comparing a population mean to a specific value such as $\mu = 100$; whereas testing the difference between two means involves comparing the means of two populations such as $\mu_1 = \mu_2$.

3.
Both samples are random samples. The populations must be independent of each other and they must be normally distributed or approximately normally distributed.

5.
$H_0: \mu_1 = \mu_2$
$H_1: \mu_1 \neq \mu_2$ (claim)
C. V. $= \pm 1.65$
$\overline{X}_1 = 8.6 \qquad \overline{X}_2 = 10.6$
$z = \dfrac{(\overline{X}_1 - \overline{X}_2) - (\mu_1 - \mu_2)}{\sqrt{\dfrac{\sigma_1^2}{n_1} + \dfrac{\sigma_2^2}{n_2}}} = \dfrac{(8.6 - 10.6) - 0}{\sqrt{\dfrac{2.1^2}{36} + \dfrac{2.7^2}{36}}}$
$z = -3.51$
(TI83 answer is $z = -3.508$)

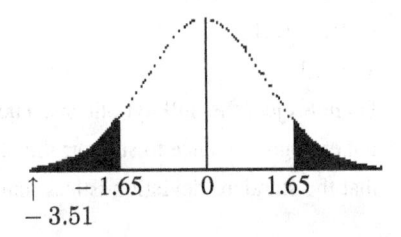

↑ −1.65 \quad 0 \quad 1.65
−3.51

Reject the null hypothesis. There is enough evidence to support the claim that the mean number of hours that families with and without children participate in recreational activities are different.

7.
$H_0: \mu_1 = \mu_2$
$H_1: \mu_1 \neq \mu_2$ (claim)
C. V. $= \pm 1.96$
$z = \dfrac{(\overline{X}_1 - \overline{X}_2) - (\mu_1 - \mu_2)}{\sqrt{\dfrac{\sigma_1^2}{n_1} + \dfrac{\sigma_2^2}{n_2}}} = \dfrac{(28.5 - 35.2) - 0}{\sqrt{\dfrac{7.2^2}{40} + \dfrac{9.1^2}{40}}}$
$z = -3.65$

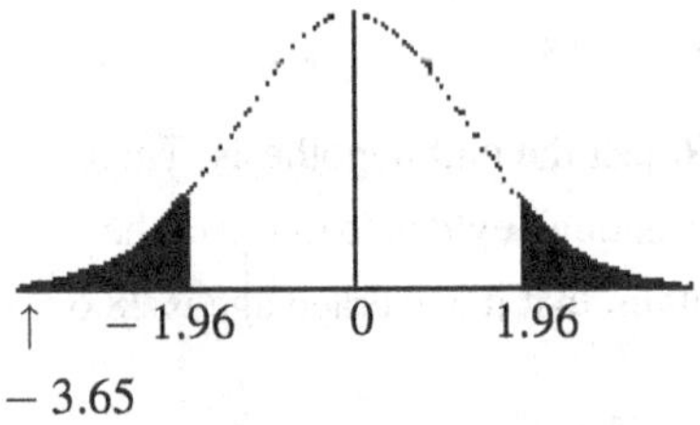

↑ −1.96 \quad 0 \quad 1.96
−3.65

Reject the null hypothesis. There is sufficient evidence at $\alpha = 0.05$ to conclude that commuting times are different in the winter.

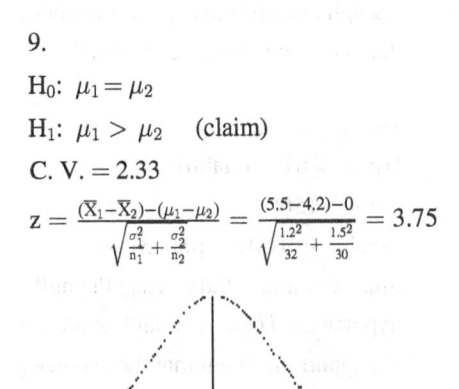

9.
$H_0: \mu_1 = \mu_2$
$H_1: \mu_1 > \mu_2$ (claim)
C. V. $= 2.33$
$z = \dfrac{(\overline{X}_1 - \overline{X}_2) - (\mu_1 - \mu_2)}{\sqrt{\dfrac{\sigma_1^2}{n_1} + \dfrac{\sigma_2^2}{n_2}}} = \dfrac{(5.5 - 4.2) - 0}{\sqrt{\dfrac{1.2^2}{32} + \dfrac{1.5^2}{30}}} = 3.75$

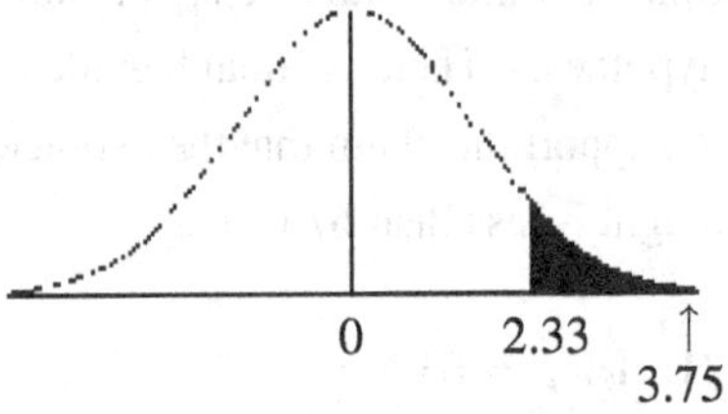

0 \quad 2.33 ↑
\qquad\qquad 3.75

Reject the null hypothesis. There is enough evidence at $\alpha = 0.01$ to support the claim that the average stay is longer for men than for women.

11.
$H_0: \mu_1 = \mu_2$
$H_1: \mu_1 \neq \mu_2$ (claim)
C. V. $= \pm 2.58$

Chapter 9 - Testing the Difference Between Two Means, Two Proportions, and Two Variances

11. continued

$$z = \frac{(\overline{X}_1 - \overline{X}_2) - (\mu_1 - \mu_2)}{\sqrt{\frac{\sigma_1^2}{n_1} + \frac{\sigma_2^2}{n_2}}} = \frac{(87-92)-0}{\sqrt{\frac{7.2^2}{30} + \frac{7.2^2}{30}}} = -2.69$$

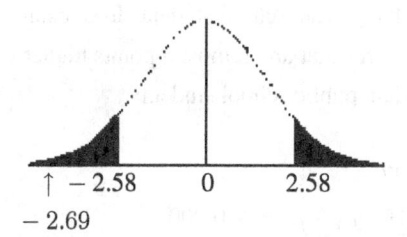

↑ −2.58   0   2.58
−2.69

Reject the null hypothesis. There is enough evidence to support the claim that the mean per capita income in Wisconsin and South Dakota are different.

13.
$H_0: \mu_1 = \mu_2$
$H_1: \mu_1 \neq \mu_2$ (claim)
C. V. = ±1.96

$$z = \frac{(\overline{X}_1 - \overline{X}_2) - (\mu_1 - \mu_2)}{\sqrt{\frac{\sigma_1^2}{n_1} + \frac{\sigma_2^2}{n_2}}}$$

$$z = \frac{(40{,}275 - 38{,}750) - 0}{\sqrt{\frac{10{,}500^2}{50} + \frac{12{,}500^2}{50}}} = 0.66$$

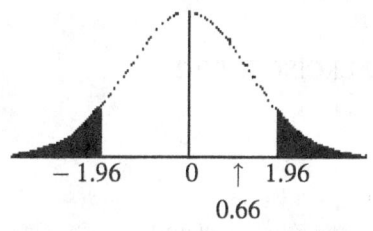

−1.96   0   ↑ 1.96
               0.66

Do not reject the null hypothesis. There is not enough evidence to support the claim that the average incomes are different.

15.
$H_0: \mu_1 = \mu_2$
$H_1: \mu_1 \neq \mu_2$ (claim)

$$z = \frac{(\overline{X}_1 - \overline{X}_2) - (\mu_1 - \mu_2)}{\sqrt{\frac{\sigma_1^2}{n_1} + \frac{\sigma_2^2}{n_2}}} = \frac{(3.05 - 2.96) - 0}{\sqrt{\frac{0.75^2}{103} + \frac{0.75^2}{225}}}$$

15. continued

$z = 1.01$
Area = 0.8438
P-value = $2(1 - 0.8438) = 0.3124$

Since P-value > 0.05, do not reject the null hypothesis. There is not enough evidence to support the claim that there is a difference in scores. (TI: P-value =0.3131)

17.
$D = 21 - 14 = 7$

$$(\overline{X}_1 - \overline{X}_2) - z_{\frac{\alpha}{2}} \sqrt{\frac{\sigma_1^2}{n_1} + \frac{\sigma_2^2}{n_2}} < \mu_1 - \mu_2 <$$
$$(\overline{X}_1 - \overline{X}_2) + z_{\frac{\alpha}{2}} \sqrt{\frac{\sigma_1^2}{n_1} + \frac{\sigma_2^2}{n_2}}$$

$$7 - (1.65)\sqrt{\frac{4.2^2}{32} + \frac{4.2^2}{32}} < \mu_1 - \mu_2 <$$
$$7 + (1.65)\sqrt{\frac{4.2^2}{32} + \frac{4.2^2}{32}}$$

$5.3 < \mu_1 - \mu_2 < 8.7$
(TI: $5.2729 < \mu_1 - \mu_2 < 8.7271$)

19.
$D = 315 - 280 = 35$

$$(\overline{X}_1 - \overline{X}_2) - z_{\frac{\alpha}{2}} \sqrt{\frac{\sigma_1^2}{n_1} + \frac{\sigma_2^2}{n_2}} < \mu_1 - \mu_2 <$$
$$(\overline{X}_1 - \overline{X}_2) + z_{\frac{\alpha}{2}} \sqrt{\frac{\sigma_1^2}{n_1} + \frac{\sigma_2^2}{n_2}}$$

$$35 - (1.96)\sqrt{\frac{56.2^2}{40} + \frac{52.1^2}{35}} < \mu_1 - \mu_2 <$$
$$35 + (1.96)\sqrt{\frac{56.2^2}{40} + \frac{52.1^2}{35}}$$

$10.5 < \mu_1 - \mu_2 < 59.5$

The interval gives evidence to reject the claim that there is no difference in the means because 0 is not contained in the interval.

21.
$H_0: \mu_1 = \mu_2$
$H_1: \mu_1 > \mu_2$ (claim)
C. V. = 2.33

21. continued

$$z = \frac{(\overline{X}_1 - \overline{X}_2) - (\mu_1 - \mu_2)}{\sqrt{\frac{\sigma_1^2}{n_1} + \frac{\sigma_2^2}{n_2}}} = \frac{(48.2 - 44.3) - 0}{\sqrt{\frac{5.6^2}{40} + \frac{4.5^2}{40}}} = 3.43$$

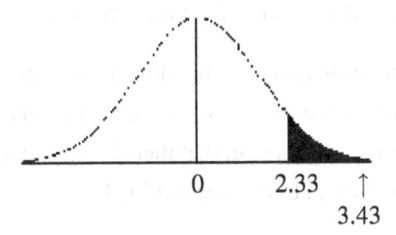

Reject the null hypothesis. There is enough evidence to support the claim that women watch more television than men.

23.

H_0: $\mu_1 = \mu_2$
H_1: $\mu_1 \neq \mu_2$ (claim)

$$z = \frac{(\overline{X}_1 - \overline{X}_2) - (\mu_1 - \mu_2)}{\sqrt{\frac{\sigma_1^2}{n_1} + \frac{\sigma_2^2}{n_2}}}$$

$$z = \frac{(\$995 - \$1120) - 0}{\sqrt{\frac{120^2}{30} + \frac{250^2}{30}}} = -2.47$$

Area = 0.0068
P-value = 2(0.0068) = 0.0136
Since P-value > 0.01, do not reject the null hypothesis. There is not enough evidence to support the claim that there is a difference in sales.

25.

H_0: $\mu_1 - \mu_2 = 8$ (claim)
H_1: $\mu_1 - \mu_2 > 8$
C. V. = 1.65

$$z = \frac{(\overline{X}_1 - \overline{X}_2) - (\mu_1 - \mu_2)}{\sqrt{\frac{\sigma_1^2}{n_1} + \frac{\sigma_2^2}{n_2}}} = \frac{(110 - 104) - 8}{\sqrt{\frac{15^2}{60} + \frac{15^2}{60}}} = -0.73$$

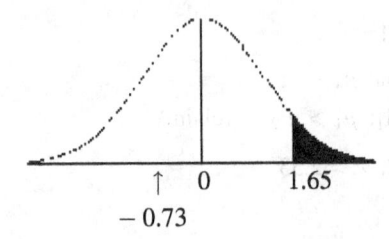

25. continued

Do not reject the null hypothesis. There is not enough evidence to reject the claim that private school students have exam scores that are at most 8 points higher than public school students.

27.

H_0: $\mu_1 - \mu_2 = \$30,000$
H_1: $\mu_1 - \mu_2 \neq \$30,000$ (claim)
C. V. = ± 2.58

$$z = \frac{(90,200 - 57,800) - 30,000}{\sqrt{\frac{15,000^2}{100} + \frac{12,800^2}{100}}} = 1.22$$

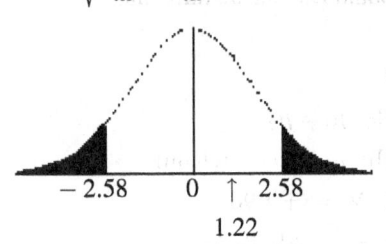

Do not reject the null hypothesis. There is not enough evidence to support the claim that the difference in income is not $30,000.

EXERCISE SET 9-2

1.

H_0: $\mu_1 = \mu_2$
H_1: $\mu_1 \neq \mu_2$ (claim)
C. V. = ± 1.860 d. f. = 8

$$t = \frac{(\overline{X}_1 - \overline{X}_2) - (\mu_1 - \mu_2)}{\sqrt{\frac{s_1^2}{n_1} + \frac{s_2^2}{n_2}}}$$

$$t = \frac{(883.22 - 840.33) - 0}{\sqrt{\frac{387.15^2}{9} + \frac{477.65^2}{9}}} = 0.209$$

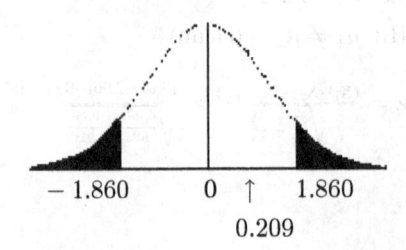

Chapter 9 - Testing the Difference Between Two Means, Two Proportions, and Two Variances

1. continued

Do not reject the null hypothesis. There is not enough evidence to support the claim that there is a significant difference in the average heights in feet of waterfalls in Europe and the ones in Asia.

3.

H_0: $\mu_1 = \mu_2$

H_1: $\mu_1 \neq \mu_2$ (claim)

C. V. $= \pm 2.093$ d. f. $= 19$

$$t = \frac{(\bar{X}_1 - \bar{X}_2) - (\mu_1 - \mu_2)}{\sqrt{\frac{s_1^2}{n_1} + \frac{s_2^2}{n_2}}}$$

$$t = \frac{(63.1 - 56.3) - 0}{\sqrt{\frac{4.1^2}{20} + \frac{7.5^2}{24}}} = 3.811$$

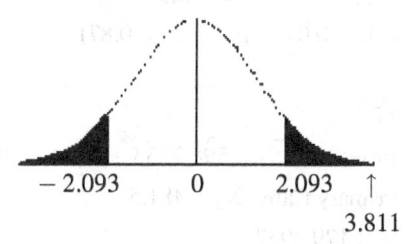

Reject the null hypothesis. There is enough evidence to support the claim that there is a difference in noise levels.

5.

H_0: $\mu_1 = \mu_2$

H_1: $\mu_1 \neq \mu_2$ (claim)

C. V. $= \pm 1.812$ d. f. $= 10$

$\bar{X}_1 = 29.69$ $s_1 = 6.499$

$\bar{X}_2 = 34.36$ $s_2 = 11.201$

$$t = \frac{(\bar{X}_1 - \bar{X}_2) - (\mu_1 - \mu_2)}{\sqrt{\frac{s_1^2}{n_1} + \frac{s_2^2}{n_2}}}$$

$$t = \frac{(29.69 - 34.36) - 0}{\sqrt{\frac{6.499^2}{13} + \frac{11.201^2}{11}}} = -1.220$$

5. continued

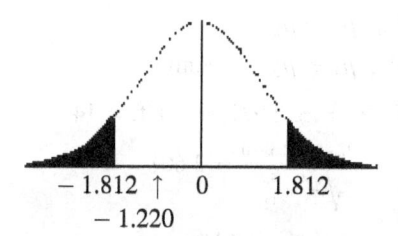

Do not reject the null hypothesis. There is not enough evidence to support the claim that there is a significant difference in the average number of grams of carbohydrates.

7.

H_0: $\mu_1 = \mu_2$

H_1: $\mu_1 \neq \mu_2$ (claim)

$\bar{X}_1 = 12.48$ $s_1 = 1.477$

$\bar{X}_2 = 9.94$ $s_2 = 0.666$

d.f. $= 9$

$$t = \frac{(\bar{X}_1 - \bar{X}_2) - (\mu_1 - \mu_2)}{\sqrt{\frac{s_1^2}{n_1} + \frac{s_2^2}{n_2}}}$$

$$t = \frac{(12.48 - 9.94) - 0}{\sqrt{\frac{1.477^2}{10} + \frac{0.666^2}{15}}} = 5.103$$

For t $= 5.103$, the area is greater than 0.9999. The P-value is $1 - 0.9999 < 0.0001$.

Since the P-value is less than $\alpha = 0.05$, reject the null hypothesis. There is enough evidence to support the claim that there is a difference in the weights of running shoes.

9.

For the 90% Confidence Interval:

$42.89 - 1.860(204.9484) < \mu_1 - \mu_2 < 42.89 + 1.860(204.9484)$

$-338.31 < \mu_1 - \mu_2 < 424.09$

(TI: $-315.90 < \mu_1 - \mu_2 < 401.65$)

Chapter 9 - Testing the Difference Between Two Means, Two Proportions, and Two Variances

11.

$H_0: \mu_1 = \mu_2$

$H_1: \mu_1 \neq \mu_2$ (claim)

C. V. = ± 2.977 d. f. = 14

$t = \dfrac{(22.45 - 18.5) - 0}{\sqrt{\dfrac{16.4}{15} + \dfrac{18.2}{15}}} = 2.601$

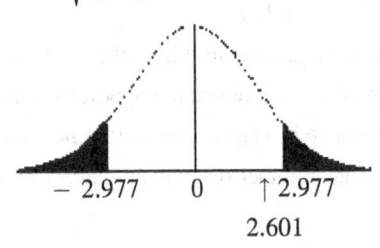

$-2.977 \quad 0 \quad \uparrow 2.977$
$\qquad\qquad\qquad 2.601$

Do not reject the null hypothesis. There is not enough evidence to support the claim that there is a difference in TV viewing times between children and teens.

13.

$H_0: \mu_1 = \mu_2$

$H_1: \mu_1 > \mu_2$ (claim)

$\overline{X}_1 = 39.6667 \quad s_1 = 18.3703$

$\overline{X}_2 = 28.8333 \quad s_2 = 17.1279$

C. V. = 3.365   d. f. = 5

$t = \dfrac{(39.6667 - 28.8333) - 0}{\sqrt{\dfrac{18.3703^2}{6} + \dfrac{17.1279^2}{6}}} = 1.057$

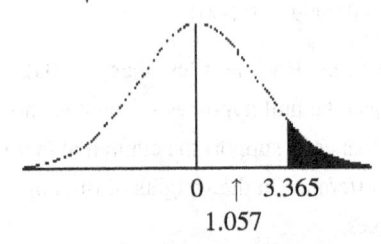

$0 \quad \uparrow 3.365$
1.057

Do not reject the null hypothesis. There is not enough evidence to support the claim that the average number of students attending cyber schools in Allegheny County is greater than those who attend cyber schools outside Allegheny County. One reason why caution should be used is that cyber charter schools are a relatively new concept.

15.

$H_0: \mu_1 = \mu_2$ (claim)

$H_1: \mu_1 \neq \mu_2$

P-value: $0.02 <$ P-value < 0.05 (0.026)

d. f. = 15

$t = \dfrac{(2.3 - 1.9) - 0}{\sqrt{\dfrac{0.6^2}{16} + \dfrac{0.3^2}{16}}} = 2.385$

Since P-value > 0.01, do not reject the null hypothesis. There is not enough evidence to reject the claim that the mean hospital stay is the same.

(TI: P-value = 0.026)

99% Confidence Interval:

$0.4 - 2.947(0.1677) < \mu_1 - \mu_2 <$
$\qquad\qquad 0.4 + 2.947(0.1677)$

$-0.1 < \mu_1 - \mu_2 < 0.9$

(TI: $-0.07 < \mu_1 - \mu_2 < 0.87$)

17.

Research: $\overline{X}_1 = 596.2353 \quad s_1 = 163.2362$

Primary Care: $\overline{X}_2 = 481.5$

$s_2 = 179.3957$

90% Confidence Interval:

$114.7353 - 1.753\sqrt{\dfrac{163.2363^2}{17} + \dfrac{179.3957^2}{16}}$

$< \mu_1 - \mu_2 <$

$114.7353 + 1.753\sqrt{\dfrac{163.2363^2}{17} + \dfrac{179.3957^2}{16}}$

$114.7353 - 104.87 < \mu_1 - \mu_2$
$\qquad\qquad < 114.7353 + 104.87$

$9.8653 < \mu_1 - \mu_2 < 219.6053$

(TI: $13.23 < \mu_1 - \mu_2 < 216.24$)

19.

$H_0: \mu_1 = \mu_2$

$H_1: \mu_1 < \mu_2$ (claim)

$\overline{X}_1 = 2.563 \quad s_1 = 0.360$

$\overline{X}_2 = 2.690 \quad s_2 = 0.223$

Chapter 9 - Testing the Difference Between Two Means, Two Proportions, and Two Variances

19. continued

$$t = \frac{(\bar{X}_1 - \bar{X}_2) - (\mu_1 - \mu_2)}{\sqrt{\frac{s_1^2}{n_1} + \frac{s_2^2}{n_2}}}$$

$$t = \frac{(2.563 - 2.690) - 0}{\sqrt{\frac{0.360^2}{6} + \frac{0.223^2}{7}}} = -0.75$$

$0.20 <$ P-value < 0.25 (0.2435). Since the P-value is greater than $\alpha = 0.01$, do not reject the null hypothesis. There is not enough evidence to support the claim that the average gasoline price in 2011 was less than the average price in 2015.

21.

H_0: $\mu_1 = \mu_2$

H_1: $\mu_1 \neq \mu_2$ (claim)

C. V. $= \pm 1.796$ d. f. $= 11$

$\bar{X}_1 = 10.17$ $s_1 = 8.943$

$\bar{X}_2 = 11.67$ $s_2 = 7.315$

$$t = \frac{(\bar{X}_1 - \bar{X}_2) - (\mu_1 - \mu_2)}{\sqrt{\frac{s_1^2}{n_1} + \frac{s_2^2}{n_2}}}$$

$$t = \frac{(10.17 - 11.67) - 0}{\sqrt{\frac{8.943^2}{12} + \frac{7.315^2}{12}}} = -0.45$$

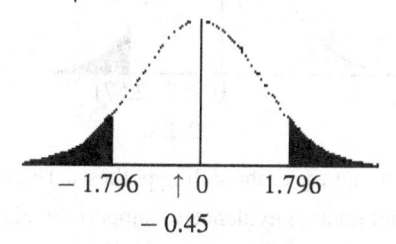

$-1.796 \uparrow 0 \quad 1.796$
$\quad -0.45$

Do not reject the null hypothesis. There is not enough evidence to support the claim that the mean number of home runs for the two leagues are different.

EXERCISE SET 9-3

1.

a. Dependent

b. Dependent

1. continued

c. Independent

d. Dependent

e. Independent

3.

Before	After	D	D²
9	9	0	0
12	17	-5	25
6	9	-3	9
15	20	-5	25
3	2	1	1
18	21	-3	9
10	15	-5	25
13	22	-9	81
7	6	1	1

$\sum D = -28$ $\sum D^2 = 176$

H_0: $\mu_D = 0$

H_1: $\mu_D < 0$ (claim)

C. V. $= -1.397$ d. f. $= 8$

$$\bar{D} = \frac{\sum D}{n} = -3.11$$

$$s_D = \sqrt{\frac{n \sum D^2 - (\sum D)^2}{n(n-1)}}$$

$$= \sqrt{\frac{9(176) - (-28)^2}{9(8)}} = 3.333$$

$$t = \frac{-3.11 - 0}{\frac{3.33}{\sqrt{9}}} = -2.818$$

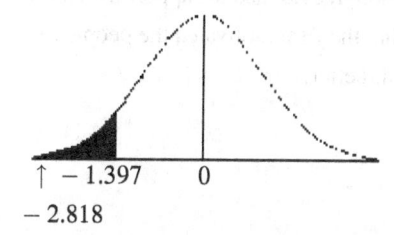

$\uparrow -1.397 \quad 0$
-2.818

Reject the null hypothesis. There is enough evidence to support the claim that the seminar increased the number of hours students studied.

Chapter 9 - Testing the Difference Between Two Means, Two Proportions, and Two Variances

5.

Before	After	D	D²
243	215	28	784
216	202	14	196
214	198	16	256
222	195	27	729
206	204	2	4
219	213	6	36
		$\sum D = 93$	$\sum D^2 = 2005$

H_0: $\mu_D = 0$
H_1: $\mu_D > 0$ (claim)

C. V. = 2.015   d. f. = 5

$\bar{D} = \frac{\sum D}{n} = \frac{93}{6} = 15.5$

$s_D = \sqrt{\frac{n \sum D^2 - (\sum D)^2}{n(n-1)}}$

$s_D = \sqrt{\frac{6(2005) - (93)^2}{6(5)}} = 10.616$

$t = \frac{15.5 - 0}{\frac{10.616}{\sqrt{6}}} = 3.58$

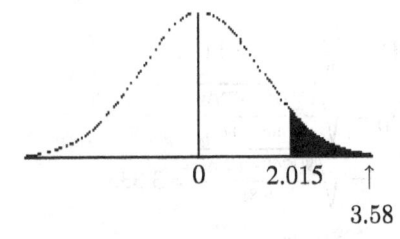

Reject the null hypothesis. There is enough evidence to support the claim that the film motivated the people to eat better.

7.

Before	After	D	D²
12	9	3	9
9	6	3	9
0	1	-1	1
5	3	2	4
4	2	2	4
3	3	0	0
		$\sum D = 9$	$\sum D^2 = 27$

H_0: $\mu_D = 0$
H_1: $\mu_D > 0$ (claim)

C. V. = 2.571   d. f. = 5

$\bar{D} = \frac{\sum D}{n} = \frac{9}{6} = 1.5$

$s_D = \sqrt{\frac{n \sum D^2 - (\sum D)^2}{n(n-1)}} = \sqrt{\frac{6(27) - 9^2}{6(5)}} = 1.643$

$t = \frac{1.5 - 0}{\frac{1.643}{\sqrt{6}}} = 2.236$

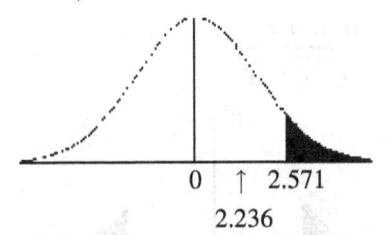

Do not reject the null hypothesis. There is not enough evidence to support the claim that the errors have been reduced.

Chapter 9 - Testing the Difference Between Two Means, Two Proportions, and Two Variances

9.

A	B	D	D^2
87	83	4	16
92	95	-3	9
78	79	-1	1
83	83	0	0
88	86	2	4
90	93	-3	9
84	80	4	16
93	86	7	49

$\sum D = 10 \quad \sum D^2 = 104$

$H_0: \mu_D = 0$
$H_1: \mu_D \neq 0$ (claim)

d. f. = 7

$\overline{D} = \frac{\sum D}{n} = \frac{10}{8} = 1.25$

$s_D = \sqrt{\frac{n \sum D^2 - (\sum D)^2}{n(n-1)}} = \sqrt{\frac{8(104) - 10^2}{8(7)}} = 3.62$

$t = \frac{1.25 - 0}{\frac{3.62}{\sqrt{8}}} = 0.978$

P-value > 0.20 (0.361). Do not reject the null hypothesis since P-value > 0.01. There is not enough evidence to support the claim that there is a difference in the pulse rates.

Confidence Interval:
$1.25 - 3.499\left(\frac{3.62}{\sqrt{8}}\right) < \mu_D <$
$\quad 1.25 + 3.499\left(\frac{3.62}{\sqrt{8}}\right)$
$-3.2 < \mu_D < 5.7$

11.

Before	After	D	D^2
8	6	2	4
3	4	-1	1
10	8	2	4
5	1	4	16
9	4	5	25
11	7	4	16
12	11	1	1

$\sum D = 17 \quad \sum D^2 = 67$

$H_0: \mu_D = 0$
$H_1: \mu_D > 0$ (claim)

C. V. = 1.943 \quad d. f. = 6

$\overline{D} = \frac{17}{7} = 2.429$

$s_D = \sqrt{\frac{7(67) - (17)^2}{7(6)}} = 2.07$

$t = \frac{2.429 - 0}{\frac{2.07}{\sqrt{7}}} = 3.104$

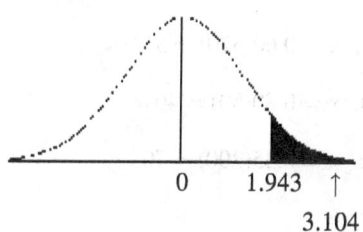

Reject the null hypothesis. There is enough evidence to support the claim that the average number of spelling errors have reduced.

13.

$\overline{X_1 - X_2} = \sum \frac{X_1 - X_2}{n}$

$\sum \frac{X_1 - X_2}{n} = \sum \left(\frac{X_1}{n} - \frac{X_2}{n}\right)$

$\sum \left(\frac{X_1}{n} - \frac{X_2}{n}\right) = \sum \frac{X_1}{n} - \sum \frac{X_2}{n}$

$\sum \frac{X_1}{n} - \sum \frac{X_2}{n} = \overline{X}_1 - \overline{X}_2$

Chapter 9 - Testing the Difference Between Two Means, Two Proportions, and Two Variances

EXERCISE SET 9-4

1

Use $\hat{p} = \dfrac{X}{n}$ and $\hat{q} = 1 - \hat{p}$

a. $\hat{p} = \dfrac{32}{52}$ $\quad$ $\hat{q} = \dfrac{20}{52}$

b. $\hat{p} = \dfrac{66}{80}$ $\quad$ $\hat{q} = \dfrac{14}{80}$

c. $\hat{p} = \dfrac{12}{36}$ $\quad$ $\hat{q} = \dfrac{24}{36}$

d. $\hat{p} = \dfrac{7}{42}$ $\quad$ $\hat{q} = \dfrac{35}{42}$

e. $\hat{p} = \dfrac{50}{160}$ $\quad$ $\hat{q} = \dfrac{110}{160}$

3.

a. $x = 0.60(240) = 144$

b. $x = 0.20(320) = 64$

c. $x = 0.60(520) = 312$

d. $x = 0.80(50) = 40$

e. $x = 0.35(200) = 70$

5.

For each part, use the formulas $\bar{p} = \dfrac{X_1 + X_2}{n_1 + n_2}$ and $\bar{q} = 1 - \bar{p}$

a. $\bar{p} = \dfrac{25 + 40}{75 + 90} = 0.3939$

$\bar{q} = 1 - 0.3939 = 0.6061$

b. $\bar{p} = \dfrac{9 + 7}{15 + 20} = 0.4571$

$\bar{q} = 1 - 0.4571 = 0.5429$

c. $\bar{p} = \dfrac{3 + 5}{20 + 40} = 0.1333$

$\bar{q} = 1 - 0.1333 = 0.8667$

5. continued

d. $\bar{p} = \dfrac{21 + 32}{50 + 50} = 0.53$

$\bar{q} = 1 - 0.53 = 0.47$

e. $\bar{p} = \dfrac{20 + 30}{150 + 50} = 0.25$

$\bar{q} = 1 - 0.25 = 0.75$

7.

$\hat{p}_1 = 0.83$ $\quad$ $\hat{p}_2 = 0.75$

$X_1 = 0.83(100) = 83$

$X_2 = 0.75(100) = 75$

$\bar{p} = \dfrac{83 + 75}{100 + 100} = 0.79$ $\quad$ $\bar{q} = 1 - 0.79 = 0.21$

H_0: $p_1 = p_2$ (claim)
H_1: $p_1 \neq p_2$

C.V. $= \pm 1.96$ $\quad$ $\alpha = 0.05$

$z = \dfrac{(\hat{p}_1 - \hat{p}_2) - (p_1 - p_2)}{\sqrt{(\bar{p})(\bar{q})\left(\dfrac{1}{n_1} + \dfrac{1}{n_2}\right)}} = \dfrac{(0.83 - 0.75) - 0}{\sqrt{(0.79)(0.21)\left(\dfrac{1}{100} + \dfrac{1}{100}\right)}}$

$z = 1.39$

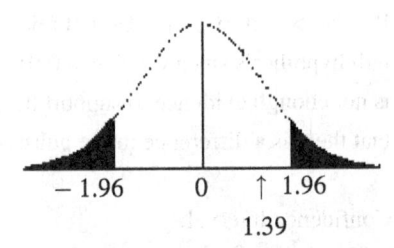

Do not reject the null hypothesis. There is not enough evidence to reject the claim that the proportions are equal.

$(\hat{p}_1 - \hat{p}_2) - z_{\frac{\alpha}{2}}\sqrt{\dfrac{\hat{p}_1 \hat{q}_1}{n_1} + \dfrac{\hat{p}_2 \hat{q}_2}{n_2}} < p_1 - p_2 <$

$(\hat{p}_1 - \hat{p}_2) + z_{\frac{\alpha}{2}}\sqrt{\dfrac{\hat{p}_1 \hat{q}_1}{n_1} + \dfrac{\hat{p}_2 \hat{q}_2}{n_2}}$

$0.08 - 1.96\sqrt{\dfrac{0.83(0.17)}{100} + \dfrac{0.75(0.25)}{100}} < p_1 - p_2$

$< 0.08 + 1.96\sqrt{\dfrac{0.83(0.17)}{100} + \dfrac{0.75(0.25)}{100}}$

$-0.032 < p_1 - p_2 < 0.192$

Chapter 9 - Testing the Difference Between Two Means, Two Proportions, and Two Variances

9.
$\hat{p}_1 = \frac{44}{80} = 0.55 \qquad \hat{p}_2 = \frac{41}{90} = 0.4556$

$\bar{p} = \frac{X_1 + X_2}{n_1 + n_2} = \frac{44 + 41}{80 + 90} = 0.5$

$\bar{q} = 1 - \bar{p} = 1 - 0.5 = 0.5$

$H_0: p_1 = p_2$
$H_1: p_1 \neq p_2$ (claim)

C. V. $= \pm 2.58 \qquad \alpha = 0.01$

$z = \frac{(\hat{p}_1 - \hat{p}_2) - (p_1 - p_2)}{\sqrt{(\bar{p})(\bar{q})\left(\frac{1}{n_1} + \frac{1}{n_2}\right)}} = \frac{(0.55 - 0.4556) - 0}{\sqrt{(0.5)(0.5)\left(\frac{1}{80} + \frac{1}{90}\right)}}$

$z = 1.23$

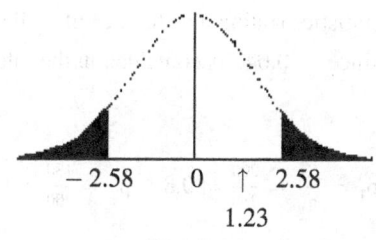

−2.58   0   ↑ 2.58
            1.23

Do not reject the null hypothesis. There is not enough evidence to support the claim that the proportions are different.

$(\hat{p}_1 - \hat{p}_2) - z_{\frac{\alpha}{2}}\sqrt{\frac{\hat{p}_1\hat{q}_1}{n_1} + \frac{\hat{p}_1\hat{q}_1}{n_2}} < p_1 - p_2 <$
$(\hat{p}_1 - \hat{p}_2) + z_{\frac{\alpha}{2}}\sqrt{\frac{\hat{p}_1\hat{q}_1}{n_1} + \frac{\hat{p}_1\hat{q}_1}{n_2}}$

$0.0944 - 2.58\sqrt{\frac{0.55(0.4556)}{80} + \frac{0.4556(0.55)}{90}}$
$< p_1 - p_2$
$< 0.0944 + 2.58\sqrt{\frac{0.55(0.4556)}{80} + \frac{0.4556(0.55)}{90}}$

$-0.104 < p_1 - p_2 < 0.293$

(TI: $-0.104 < p_1 - p_2 < 0.293$)

11.
$\hat{p}_1 = \frac{14}{50} = 0.28 \qquad \hat{p}_2 = \frac{21}{60} = 0.35$

$\bar{p} = \frac{X_1 + X_2}{n_1 + n_2} = \frac{14 + 21}{50 + 60} = 0.3182$

$q = 1 - \bar{p} = 1 - 0.3852 = 0.6818$

$H_0: p_1 = p_2$
$H_1: p_1 < p_2$ (claim)

11. continued

C. V. $= -1.65 \qquad \alpha = 0.05$

$z = \frac{(\hat{p}_1 - \hat{p}_2) - (p_1 - p_2)}{\sqrt{(\bar{p})(\bar{q})\left(\frac{1}{n_1} + \frac{1}{n_2}\right)}} = \frac{(0.28 - 0.35) - 0}{\sqrt{(0.3182)(0.6818)\left(\frac{1}{50} + \frac{1}{60}\right)}}$

$z = -0.78$

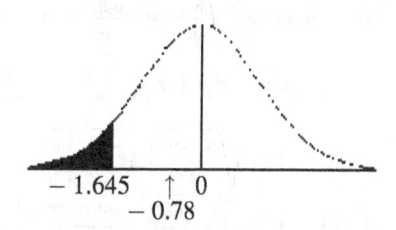

−1.645 ↑   0
    −0.78

Do not reject the null hypothesis. There is not enough evidence to support the claim that less household owners have cats than household owners who have dogs as pets.

13.
$\hat{p}_1 = \frac{X_1}{n_1} = \frac{24}{80} = 0.30$

$\hat{p}_2 = \frac{X_2}{n_2} = \frac{6}{50} = 0.12$

$\bar{p} = \frac{X_1 + X_2}{n_1 + n_2} = \frac{24 + 6}{80 + 50} = 0.2308$

$\bar{q} = 1 - \bar{p} = 1 - 0.2308 = 0.7692$

$H_0: p_1 = p_2$
$H_1: p_1 > p_2$ (claim)

C. V. $= 1.28 \qquad \alpha = 0.10$

$z = \frac{(\hat{p}_1 - \hat{p}_2) - (p_1 - p_2)}{\sqrt{(\bar{p})(\bar{q})\left(\frac{1}{n_1} + \frac{1}{n_2}\right)}} = \frac{(0.30 - 0.12) - 0}{\sqrt{(0.2308)(0.7692)\left(\frac{1}{80} + \frac{1}{50}\right)}}$

$z = 2.37$

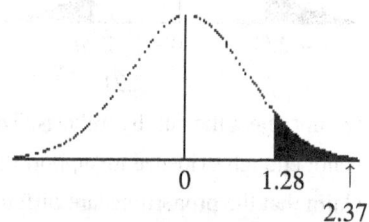

0   1.28 ↑
          2.37

Reject the null hypothesis. There is enough evidence to support the claim that the percentage of women who were attacked by relatives is greater than the percentage of men who were attacked by relatives.

15.

$\alpha = 0.05$
$\hat{p}_1 = 0.8$ $\qquad \hat{q}_1 = 0.20$
$\hat{p}_1 = 0.517$ $\qquad \hat{q}_1 = 0.483$

$\hat{p}_1 - \hat{p}_2 = 0.8 - 0.517 = 0.283$

$(\hat{p}_1 - \hat{p}_2) - z_{\frac{\alpha}{2}}\sqrt{\frac{\hat{p}_1\hat{q}_1}{n_1} + \frac{\hat{p}_1\hat{q}_1}{n_2}} < p_1 - p_2 <$
$\qquad (\hat{p}_1 - \hat{p}_2) + z_{\frac{\alpha}{2}}\sqrt{\frac{\hat{p}_1\hat{q}_1}{n_1} + \frac{\hat{p}_1\hat{q}_1}{n_2}}$

$0.283 - 1.96\sqrt{\frac{(0.8)(0.2)}{100} + \frac{(0.517)(0.483)}{120}} < p_1 - p_2 <$
$0.283 + 1.96\sqrt{\frac{(0.8)(0.2)}{100} + \frac{(0.517)(0.483)}{120}}$

$0.164 < p_1 - p_2 < 0.402$

17.

$\hat{p}_1 = \frac{80}{200} = 0.4 \quad \hat{p}_2 = \frac{59}{200} = 0.295$

$\bar{p} = \frac{X_1 + X_2}{n_1 + n_2} = \frac{80 + 59}{200 + 200} = 0.3475$

$\bar{q} = 1 - \bar{p} = 1 - 0.3475 = 0.6525$

H_0: $p_1 = p_2$
H_1: $p_1 \neq p_2$ (claim)

C. V. $= \pm 2.58$

$z = \frac{(\hat{p}_1 - \hat{p}_2) - (p_1 - p_2)}{\sqrt{(\bar{p})(\bar{q})(\frac{1}{n_1} + \frac{1}{n_2})}} = \frac{(0.4 - 0.295) - 0}{\sqrt{(0.3475)(0.6525)(\frac{1}{200} + \frac{1}{200})}}$

$z = 2.21$

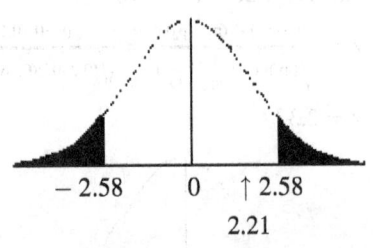

$-2.58 \qquad 0 \qquad \uparrow 2.58$
$\qquad\qquad\qquad 2.21$

Do not reject the null hypothesis. There is not enough evidence to support the claim that the proportions are different.

19.

$\hat{p}_1 = \frac{100}{350} = 0.2857 \qquad \hat{q}_1 = 0.7143$

$\hat{p}_2 = \frac{115}{400} = 0.2875 \qquad \hat{q}_2 = 0.7125$

19. continued

$\hat{p}_1 - \hat{p}_2 = -0.0018$

$(\hat{p}_1 - \hat{p}_2) - z_{\frac{\alpha}{2}}\sqrt{\frac{\hat{p}_1\hat{q}_1}{n_1} + \frac{\hat{p}_1\hat{q}_1}{n_2}} < p_1 - p_2 <$
$\qquad (\hat{p}_1 - \hat{p}_2) + z_{\frac{\alpha}{2}}\sqrt{\frac{\hat{p}_1\hat{q}_1}{n_1} + \frac{\hat{p}_1\hat{q}_1}{n_2}}$

$-0.0018 - 1.96\sqrt{\frac{(0.2857)(0.7143)}{350} + \frac{(0.2875)(0.7125)}{400}}$
$< p_1 - p_2 <$
$-0.0018 + 1.96\sqrt{\frac{(0.2857)(0.7143)}{350} + \frac{(0.2875)(0.7125)}{400}}$

$-0.0667 < p_1 - p_2 < 0.0631$

The interval does agree with the *Almanac* statistics stating a difference of -0.042 since -0.042 is contained in the interval.

21.

$\hat{p}_1 = \frac{X_1}{n_1} = \frac{200}{250} = 0.8 \quad \hat{p}_2 = \frac{180}{300} = 0.6$

$\bar{p} = \frac{X_1 + X_2}{n_1 + n_2} = \frac{200 + 180}{250 + 300} = 0.691$

$\bar{q} = 1 - \bar{p} = 1 - 0.691 = 0.31$

H_0: $p_1 = p_2$
H_1: $p_1 \neq p_2$ (claim)

C. V. $= \pm 2.58$

$z = \frac{(\hat{p}_1 - \hat{p}_2) - (p_1 - p_2)}{\sqrt{(\bar{p})(\bar{q})(\frac{1}{n_1} + \frac{1}{n_2})}} = \frac{(0.8 - 0.6) - 0}{\sqrt{(0.691)(0.31)(\frac{1}{250} + \frac{1}{300})}}$

$z = 5.05$

$-2.58 \qquad 0 \qquad 2.58 \uparrow$
$\qquad\qquad\qquad\qquad 5.05$

Reject the null hypothesis. There is enough evidence to support the claim that the proportion of students receiving aid has changed.

23.

$\hat{p}_1 = \frac{X_1}{n_1} = \frac{72}{120} = 0.6$

$\hat{p}_2 = \frac{80}{150} = 0.533$

$\bar{p} = \frac{X_1 + X_2}{n_1 + n_2} = \frac{72 + 80}{120 + 150} = 0.563$

$\bar{q} = 1 - \bar{p} = 1 - 0.563 = 0.437$

H_0: $p_1 = p_2$

H_1: $p_1 \neq p_2$ (claim)

C. V. $= \pm 2.58$

$z = \frac{(\hat{p}_1 - \hat{p}_2) - (p_1 - p_2)}{\sqrt{(\bar{p})(\bar{q})\left(\frac{1}{n_1} + \frac{1}{n_2}\right)}} = \frac{(0.6 - 0.533) - 0}{\sqrt{(0.563)(0.437)\left(\frac{1}{120} + \frac{1}{150}\right)}}$

$z = 1.10$

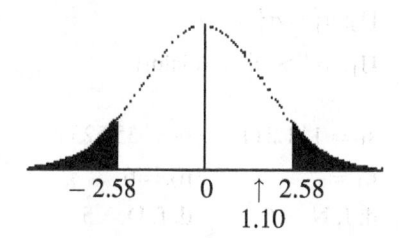

Do not reject the null hypothesis. There is not enough evidence to support the claim that the proportions are different between male interviewees and female interviewees.

25.

$\hat{p}_1 = \frac{132}{180} = 0.733$ $\hat{p}_2 = \frac{56}{100} = 0.56$

$\bar{p} = \frac{X_1 + X_2}{n_1 + n_2} = \frac{132 + 56}{180 + 100} = \frac{188}{280} = 0.671$

$\bar{q} = 1 - \bar{p} = 1 - 0.671 = 0.329$

H_0: $p_1 = p_2$

H_1: $p_1 > p_2$ (claim)

$z = \frac{(\hat{p}_1 - \hat{p}_2) - (p_1 - p_2)}{\sqrt{(\bar{p})(\bar{q})\left(\frac{1}{n_1} + \frac{1}{n_2}\right)}} = \frac{(0.733 - 0.56) - 0}{\sqrt{(0.671)(0.329)\left(\frac{1}{180} + \frac{1}{100}\right)}}$

$z = 2.96$

(TI: $z = 2.9589$; P-value $= 0.00154$)

P-value < 0.002 (0.0015).

25. continued

Since P-value $< \alpha$, reject the null hypothesis. There is enough evidence to support the claim that the proportion of women who use coupons is greater than the proportion of men who use coupons.

27.

$\hat{p}_1 = \frac{13}{200} = 0.065$ $\hat{p}_2 = \frac{16}{200} = 0.08$

$\bar{p} = \frac{13 + 16}{200 + 200} = 0.0725$

$\bar{q} = 1 - \bar{p} = 1 - 0.0725 = 0.9275$

H_0: $p_1 = p_2$

H_1: $p_1 \neq p_2$ (claim)

C. V. $= \pm 1.96$

$z = \frac{(0.065 - 0.08) - 0}{\sqrt{(0.0725)(0.9275)\left(\frac{1}{200} + \frac{1}{200}\right)}} = -0.58$

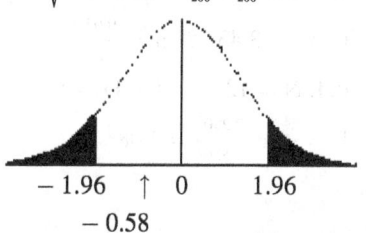

Do not reject the null hypothesis. There is not enough evidence to support the claim that there is a difference in the proportions.

EXERCISE SET 9-5

1.
The variance in the numerator should be the larger of the two variances.

3.
One degree of freedom is used for the variance associated with the numerator, and one is used for the variance associated with the denominator.

Chapter 9 - Testing the Difference Between Two Means, Two Proportions, and Two Variances

5.
a. d. f. N = 24, d. f. D = 13; C. V. = 2.89

b. d. f. N = 15, d. f. D = 11; C. V. = 2.17

c. d. f. N = 20, d. f. D = 17; C. V. = 3.16

7.
Note: Specific P-values are in parentheses.

a. $0.025 <$ P-value < 0.05 (0.033)

b. $0.05 <$ P-value < 0.10 (0.072)

c. P-value $= 0.05$

d. $0.005 <$ P-value < 0.01 (0.006)

9.
H_0: $\sigma_1^2 = \sigma_2^2$
H_1: $\sigma_1^2 \neq \sigma_2^2$ (claim)

$s_1 = 2.290$ $s_2 = 1.586$
C. V. $= 3.43$ $\alpha = \frac{0.05}{2}$
d. f. N $= 12$ d. f. D $= 11$
$F = \frac{s_1^2}{s_2^2} = \frac{2.290^2}{1.586^2} = 2.08$

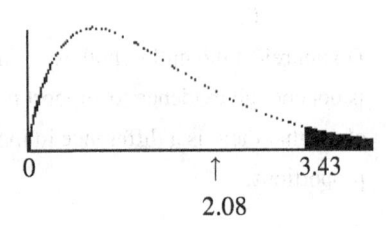

Do not reject the null hypothesis. There is not enough evidence to support the claim that the variances are different.

11.
H_0: $\sigma_1^2 = \sigma_2^2$
H_1: $\sigma_1^2 \neq \sigma_2^2$ (claim)

$s_1 = 33.99$ $s_2 = 33.99$
C. V. $= 4.99$ $\alpha = \frac{0.05}{2}$
d. f. N $= 7$ d. f. D $= 7$

11. continued

$F = \frac{s_1^2}{s_2^2} = \frac{(33.99)^2}{(33.99)^2} = 1$

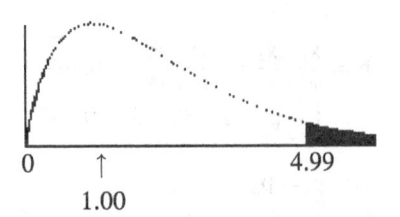

Do not reject the null hypothesis. There is not enough evidence to support the claim that the variances are different.

13.
H_0: $\sigma_1^2 = \sigma_2^2$
H_1: $\sigma_1^2 > \sigma_2^2$ (claim)

$s_1 = 111.211$ $s_2 = 35.523$
$n_1 = 7$ $n_2 = 6$
d. f. N $= 6$ d. f. D $= 5$
C. V. $= 4.950$ at $\alpha = 0.05$
C. V. $= 10.67$ at $\alpha = 0.01$

$F = \frac{s_1^2}{s_2^2} = \frac{(111.211)^2}{(35.523)^2} = 9.80$

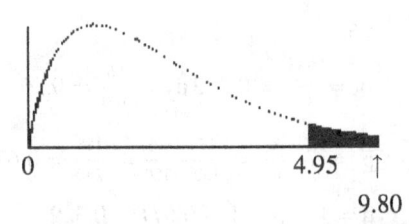

Reject the null hypothesis at $\alpha = 0.05$. There is enough evidence to support the claim that the variance in area is greater for Eastern cities.

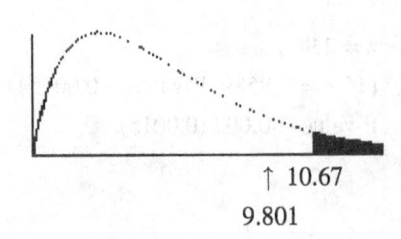

Chapter 9 - Testing the Difference Between Two Means, Two Proportions, and Two Variances

13. continued

Do not reject the null hypothesis at $\alpha = 0.01$. There is not enough evidence to support the claim that the variance in area is greater for Eastern cities.

15.

H_0: $\sigma_1^2 = \sigma_2^2$

H_1: $\sigma_1^2 \neq \sigma_2^2$ (claim)

Research: $s_1 = 5501.118$

Primary Care: $s_2 = 5238.809$

C. V. $= 4.03$ $\alpha = \frac{0.05}{2}$

d. f. N $= 9$ d. f. D $= 9$

$F = \frac{s_1^2}{s_2^2} = \frac{(5501.118)^2}{(5238.809)^2} = 1.10$

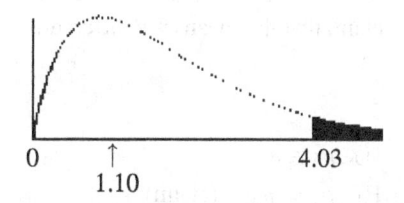

Do not reject the null hypothesis. There is not enough evidence to support the claim that there is a difference between the variances in tuition costs.

17.

H_0: $\sigma_1^2 = \sigma_2^2$ (claim)

H_1: $\sigma_1^2 \neq \sigma_2^2$

$s_1 = 130.496$ $s_2 = 71.753$

C. V. $= 3.87$ $\alpha = \frac{0.10}{2}$

d. f. N $= 6$ d. f. D $= 7$

$F = \frac{s_1^2}{s_2^2} = \frac{(130.496)^2}{(71.753)^2} = 3.31$

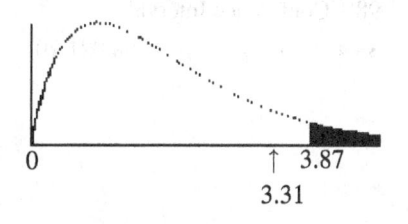

17. continued

Do not reject the null hypothesis. There is not enough evidence to reject the claim that the variances of the heights are equal.

19.

Men	Women
$s_1^2 = 2.363$	$s_2^2 = 0.444$
$n_1 = 15$	$n_2 = 15$

H_0: $\sigma_1^2 = \sigma_2^2$ (claim)

H_1: $\sigma_1^2 \neq \sigma_2$

$\alpha = 0.05$ P-value $= 0.004$

d. f. N $= 14$ d. f. D $= 14$

$F = \frac{s_1^2}{s_2^2} = \frac{2.363}{0.444} = 5.32$

Since P-value $< \alpha$, reject the null hypothesis. There is enough evidence to reject the claim that the variances in weights are equal.

21.

H_0: $\sigma_1^2 = \sigma_2^2$

H_1: $\sigma_1^2 \neq \sigma_2$ (claim)

C. V. $= 4.67$ $\alpha = \frac{0.05}{2}$

d. f. N $= 12$ d. f. D $= 7$

$F = \frac{1.3^2}{0.7^2} = 3.45$

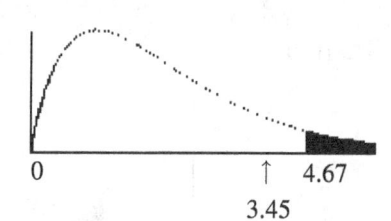

Do not reject the null hypothesis. There is not enough evidence to support the claim that the variances of ages of dog owners in Miami and Boston are different.

23.
H_0: $\sigma_1^2 = \sigma_2^2$
H_1: $\sigma_1^2 > \sigma_2^2$ (claim)
C. V. = 2.54 $\alpha = 0.05$
d. f. N = 10   d. f. D = 15
$F = \frac{s_1^2}{s_2^2} = \frac{3.2^2}{2.8^2} = 1.31$

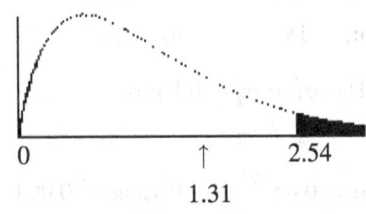

Do not reject the null hypothesis. There is not enough evidence to support the claim that the variance of the final exam scores for the students who took the online course is greater than the variance of the final exam scores of the students who took the classroom final exam.

REVIEW EXERCISES - CHAPTER 9

1.
H_0: $\mu_1 = \mu_2$
H_1: $\mu_1 > \mu_2$ (claim)
CV = 2.33 $\alpha = 0.01$
$\bar{X}_1 = 120.1$ $\bar{X}_2 = 117.8$
$s_1 = 16.722$ $s_2 = 16.053$
$z = \frac{(\bar{X}_1 - \bar{X}_2) - (\mu_1 - \mu_2)}{\sqrt{\frac{\sigma_1^2}{n_1} + \frac{\sigma_2^2}{n_2}}} = \frac{(120.1 - 117.8) - 0}{\sqrt{\frac{16.722^2}{36} + \frac{16.053^2}{35}}}$
$z = 0.59$

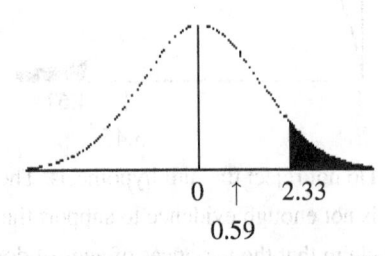

Do not reject the null hypothesis. There is not enough evidence to support the claim that single people do more pleasure driving than married people.

3.
H_0: $\mu_1 = \mu_2$
H_1: $\mu_1 \neq \mu_2$ (claim)
C. V. = ± 2.861 d. f. = 19
$\bar{X}_1 = 9.6$ $\bar{X}_2 = 10.3$
$t = \frac{(\bar{X}_1 - \bar{X}_2) - (\mu_1 - \mu_2)}{\sqrt{\frac{s_1^2}{n_1} + \frac{s_2^2}{n_2}}} = \frac{(9.6 - 10.3) - 0}{\sqrt{\frac{2.8^2}{20} + \frac{2.3^2}{25}}}$
$t = -0.901$

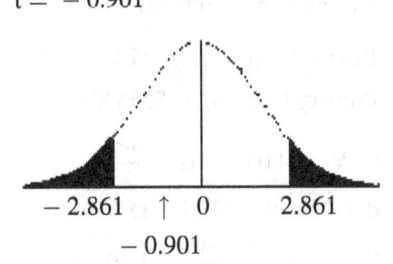

Do not reject the null hypothesis. There is not enough evidence to support the claim that the means are different.

5.
H_0: $\mu_1 = \mu_2$
H_1: $\mu_1 \neq \mu_2$ (claim)
C. V. = ± 2.624 d. f. = 14
$t = \frac{(\bar{X}_1 - \bar{X}_2) - (\mu_1 - \mu_2)}{\sqrt{\frac{s_1^2}{n_1} + \frac{s_2^2}{n_2}}} = \frac{(35,270 - 29,512) - 0}{\sqrt{\frac{3256^2}{15} + \frac{1432^2}{15}}}$
$t = 6.540$

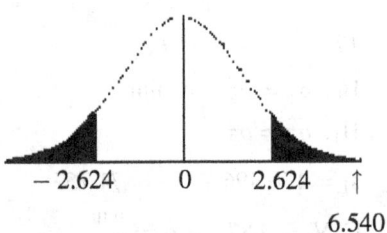

Reject the null hypothesis. There is enough evidence to support the claim that there is a difference in the teachers' salaries.

98% Confidence Interval:
$\$3,494.80 < \mu_1 - \mu_2 < \$8,021.20$

Chapter 9 - Testing the Difference Between Two Means, Two Proportions, and Two Variances

7.

Maximum	Minimum	D	D^2
44	27	17	289
46	34	12	144
46	24	22	484
36	19	17	289
34	19	15	225
36	26	10	100
57	33	24	576
62	57	5	25
73	46	27	729
53	26	27	729
		$\sum D = 176$	$\sum D^2 = 3590$

H_0: $\mu_D = 10$

H_1: $\mu_D > 10$

C. V. = 2.821

$\overline{D} = \frac{176}{10} = 17.6$

$s_D = \sqrt{\frac{10(3590) - 176^2}{10(9)}} = 7.3967$

$t = \frac{17.6 - 10}{\frac{7.3967}{\sqrt{10}}} = 3.249$

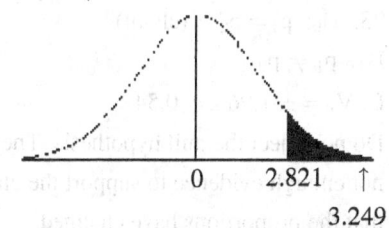

Reject the null hypothesis. There is enough evidence to support the claim that the mean difference is more than 10 degrees.

9.

$\hat{p}_1 = \frac{49}{200} = 0.245$ $\hat{p}_2 = \frac{62}{200} = 0.31$

$\overline{p} = \frac{49 + 62}{200 + 200} = 0.2775$

$\overline{q} = 1 - 0.2775 = 0.7225$

H_0: $p_1 = p_2$

H_1: $p_1 \neq p_2$ (claim)

9. continued

C. V. = ± 1.96

$z = \frac{(0.245 - 0.31) - 0}{\sqrt{(0.2775)(0.7225)(\frac{1}{200} + \frac{1}{200})}} = -1.45$

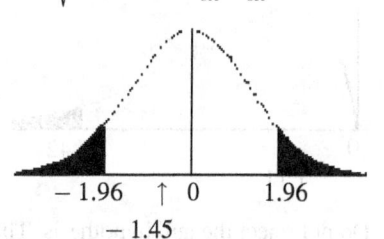

Do not reject the null hypothesis. There is not enough evidence to support the claim that the proportions of men and women who gamble are different.

11.

H_0: $\sigma_1 = \sigma_2$

H_1: $\sigma_1 \neq \sigma_2$ (claim)

$\alpha = 0.10$

d.f.N. = 23   d.f.D. = 10

C. V. = 2.77

$F = \frac{13.2^2}{4.1^2} = 10.37$

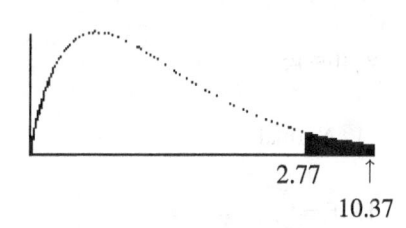

Reject the null hypothesis. There is enough evidence to support the claim that there is a difference in the standard deviations.

13.

H_0: $\sigma_1^2 = \sigma_2^2$

H_1: $\sigma_1^2 \neq \sigma_2^2$ (claim)

C. V. = 5.42 $\alpha = \frac{0.01}{2}$

d.f.N. = 11   d.f.D. = 11

Chapter 9 - Testing the Difference Between Two Means, Two Proportions, and Two Variances

13. continued

$$F = \frac{4.868^2}{4.619^2} = 1.11$$

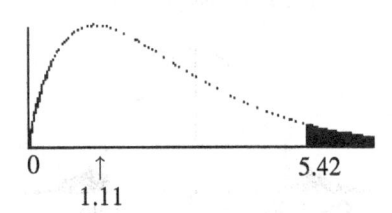

Do not reject the null hypothesis. There is not enough evidence to support the claim that there is a difference in the variances.

CHAPTER 9 QUIZ

1. False, there are different formulas for independent and dependent samples.

3. True

5. d

7. c

9. $\mu_1 = \mu_2$

11. Normal

13. $F = \dfrac{s_1^2}{s_2^2}$

15. H_0: $\mu_1 = \mu_2$
H_1: $\mu_1 > \mu_2$ (claim)
C. V. = 1.28   z = 1.61
Reject the null hypothesis. There is enough evidence to support the claim that average rental fees for the Eastern apartments is greater than the average rental fees for the Western apartments.

17. H_0: $\mu_1 = \mu_2$
H_1: $\mu_1 < \mu_2$ (claim)
C. V. = −2.132 d.f. = 4 t = −4.05 Reject the null hypothesis. There is enough evidence to support the claim that accidents have increased.

19. H_0: $\mu_1 = \mu_2$
H_1 $\mu_1 > \mu_2$ (claim)
d. f. = 10   = 0.874
0.10 < P-value < 0.25 (0.198)
Do not reject the null hypothesis since P-value > 0.05. There is not enough evidence to support the claim that the incomes of city residents are greater than the incomes of rural residents.

21. H_0: $\mu_D = 0$
H_1 $\mu_D < 0$ (claim)
C. V. = −1.833   t = −1.714
Do not reject the null hypothesis. There is not enough evidence to support the claim that egg production was increased.

23. H_0: $p_1 = p_2$ (claim)
H_1: $p_1 \neq p_2$
C. V. = ±1.96 z = 0.54
Do not reject the null hypothesis. There is not enough evidence to support the claim that the proportions have changed.

95% Confidence Interval:
−0.026 < $p_1 - p_2$ < 0.0460

Yes, the confidence interval contains 0; hence, the null hypothesis is not rejected.

25. H_0: $\sigma_1^2 = \sigma_2^2$
H_1 $\sigma_1^2 \neq \sigma_2^2$
C. V. = 1.90   F = 1.30
Do not reject. There is not enough evidence to support the claim that the variances are different.

Chapter 10 - Correlation and Regression

Note: Graphs are not to scale and are intended to convey a general idea.

Answers may vary due to rounding, TI-83's, or computer programs.

EXERCISE SET 10-1

1.
Two variables are related when there exists a discernible pattern between them.

3.
r, ρ (rho)

5.
A positive relationship means that as x increases, y also increases.
A negative relationship means that as x increases, y decreases.

7.
The diagram is called a scatter plot. It shows the nature of the relationship.

9.
t test

11.

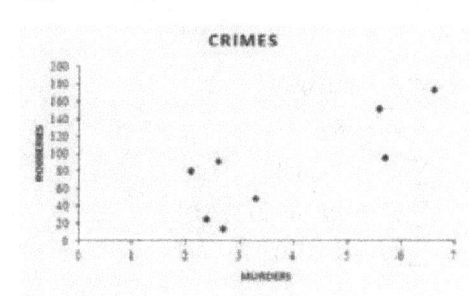

$\sum x = 31$
$\sum y = 680.1$
$\sum x^2 = 142.52$
$\sum y^2 = 80,033.99$
$\sum xy = 3202.71$
n = 8

11. continued

$$r = \frac{n(\sum xy)-(\sum x)(\sum y)}{\sqrt{[n(\sum x^2)-(\sum x)^2][n(\sum y^2)-(\sum y)^2]}}$$

$$r = \frac{8(3202.71)-(31)(680.1)}{\sqrt{[8(142.52)-31^2][8(80,033.99)-680.1^2]}}$$

$r = 0.804$

$H_0: \rho = 0$
$H_1: \rho \neq 0$

C. V. = ± 0.707 d. f. = 6
Decision: Reject. There is a significant linear relationship between the number of murders and the number of robberies per 100,000 people in a random sample of states.

13.

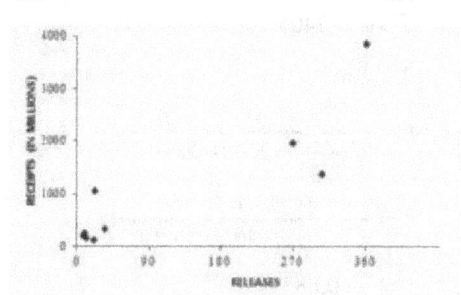

$\sum x = 1045$
$\sum y = 9283$
$\sum x^2 = 299,315$
$\sum y^2 = 21,881,839$
$\sum xy = 2,380,435$
n = 9

$$r = \frac{n(\sum xy)-(\sum x)(\sum y)}{\sqrt{[n(\sum x^2)-(\sum x)^2][n(\sum y^2)-(\sum y)^2]}}$$

$$r = \frac{9(2,380,435)-(1045)(9283)}{\sqrt{[9(299,315)-1045^2][9(21,881,839)-9283^2]}}$$

$r = 0.880$

$H_0: \rho = 0$
$H_1: \rho \neq 0$

C. V. = ± 0.666 d. f. = 7

Chapter 10 - Correlation and Regression

13. continued

Decision: Reject. There is a significant linear relationship between number of movie releases and gross receipts.

15.

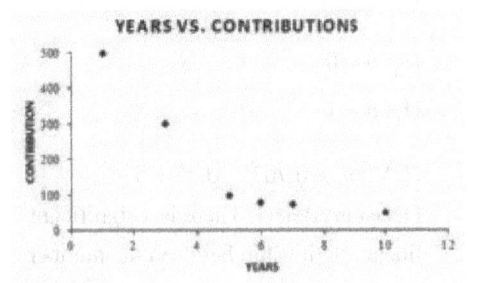

YEARS VS. CONTRIBUTIONS

$\sum x = 32$
$\sum y = 1105$
$\sum x^2 = 220$
$\sum y^2 = 364,525$
$\sum xy = 3405$
n = 6

$r = \dfrac{n(\sum xy)-(\sum x)(\sum y)}{\sqrt{[n(\sum x^2)-(\sum x)^2][n(\sum y^2)-(\sum y)^2]}}$

$r = \dfrac{6(3405)-(32)(1105)}{\sqrt{[6(220)-32^2][6(364525)-1105^2]}}$

$r = -0.883$

H_0: $\rho = 0$
H_1: $\rho \neq 0$

C. V. = ± 0.811 d. f. = 4

Decision: Reject. There is a significant linear relationship between the number of years a person has been out of school and his or her contributions.

17.

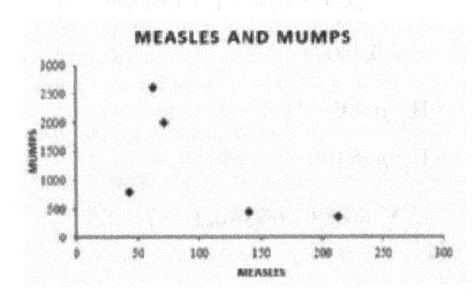

MEASLES AND MUMPS

17. continued

$\sum x = 529$
$\sum y = 6227$
$\sum x^2 = 75,403$
$\sum y^2 = 11,769,641$
$\sum xy = 482,317$
n = 5

$r = \dfrac{n(\sum xy)-(\sum x)(\sum y)}{\sqrt{[n(\sum x^2)-(\sum x)^2][n(\sum y^2)-(\sum y)^2]}}$

$r = \dfrac{5(482,317)-(529)(6227)}{\sqrt{[5(75,403)-529^2][5(11,769,641)-6227^2]}}$

$r = -0.632$

H_0: $\rho = 0$
H_1: $\rho \neq 0$

C. V. = ± 0.878 d. f. = 3

Decision: Do not reject. There is no significant linear relationship between the number of cases of measles and mumps.

19.

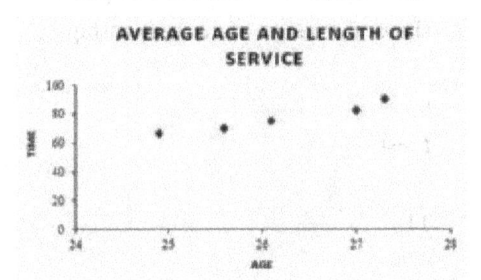

AVERAGE AGE AND LENGTH OF SERVICE

$\sum x = 130.9$
$\sum y = 383.5$
$\sum x^2 = 3430.87$
$\sum y^2 = 29,768.21$
$\sum xy = 10,076.41$
n = 5

$r = \dfrac{5(10,076.41)-(130.9)(383.5)}{\sqrt{[5(3430.87)-130.9^2][5(29,768.21)-383.5^2]}}$

$r = 0.978$

H_0: $\rho = 0$
H_1: $\rho \neq 0$

Chapter 10 - Correlation and Regression

19. continued

C. V. = ± 0.878   d. f. = 3

Decision: Reject. There is a significant linear relationship between the average age and length of service in months.

21.

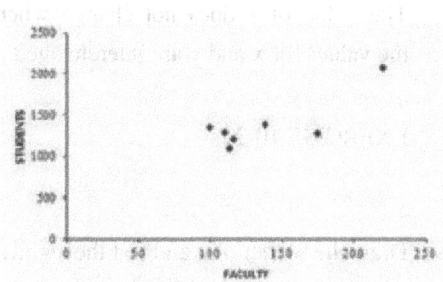

$\sum x = 970$

$\sum y = 9689$

$\sum x^2 = 145,846$

$\sum y^2 = 14,023,529$

$\sum xy = 1,410,572$

n = 7

$r = \dfrac{n(\sum xy)-(\sum x)(\sum y)}{\sqrt{[n(\sum x^2)-(\sum x)^2][n(\sum y^2)-(\sum y)^2]}}$

$r = \dfrac{7(1,410,572)-(970)(9689)}{\sqrt{[7(145,846)-970^2][7(14,023,529)-9689^2]}}$

$r = 0.812$

$H_0: \rho = 0$

$H_1: \rho \neq 0$

C. V. = ± 0.754   d. f. = 5

Decision: Reject. There is a significant linear relationship between the number of faculty and the number of students.

For x = Students and y = Faculty:

$r = \dfrac{7(1,410,572)-(9689)(970)}{\sqrt{[7(14,023,529)-9689^2][7(145,846)-970^2]}}$

$r = 0.812$

The results are the same when x and y are switched. Students would be the more likely independent variable.

23.

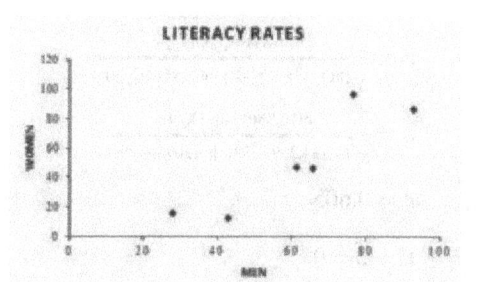

$\sum x = 367.5$

$\sum y = 302.7$

$\sum x^2 = 25,192.41$

$\sum y^2 = 21,346.59$

$\sum xy = 22,207.31$

n = 6

$r = \dfrac{n(\sum xy)-(\sum x)(\sum y)}{\sqrt{[n(\sum x^2)-(\sum x)^2][n(\sum y^2)-(\sum y)^2]}}$

$r = \dfrac{6(22,207.31)-(367.5)(302.7)}{\sqrt{[6(25,192.41)-367.5^2][6(21,346.59)-302.7^2]}}$

$r = 0.908$

$H_0: \rho = 0$

$H_1: \rho \neq 0$

C. V. = ± 0.811   d. f. = 4

Decision: Reject. There is a significant linear relationship between the literacy rates of men and the literacy rates of women.

25.

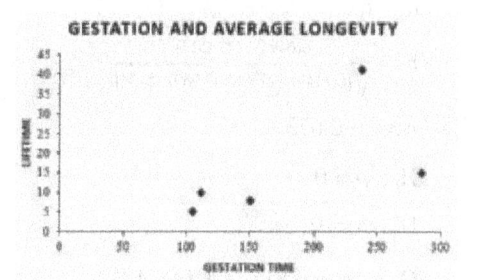

$\sum x = 891$

$\sum y = 79$

$\sum x^2 = 184,239$

$\sum y^2 = 2095$

$\sum xy = 16,886$

n = 5

25. continued

$$r = \frac{n(\sum xy)-(\sum x)(\sum y)}{\sqrt{[n(\sum x^2)-(\sum x)^2][n(\sum y^2)-(\sum y)^2]}}$$

$$r = \frac{5(16{,}886)-(891)(79)}{\sqrt{[5(184{,}239)-891^2][5(2095)-79^2]}}$$

$r = 0.605$

$H_0: \rho = 0$
$H_1: \rho \neq 0$

C. V. $= \pm 0.878$ d. f. $= 3$

Decision: Do not reject. There is not enough evidence to support the claim that there is a significant relationship between the gestation time and the lifetime.

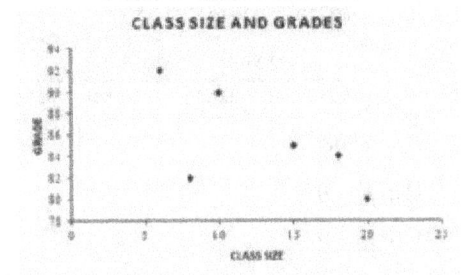

27.

CLASS SIZE AND GRADES

$\sum x = 77$
$\sum y = 513$
$\sum x^2 = 1149$
$\sum y^2 = 43{,}969$
$\sum xy = 6495$
n = 6

$$r = \frac{6(6495)-(77)(513)}{\sqrt{[6(1149)-77^2][6(43{,}969)-513^2]}}$$

$r = -0.673$

$H_0: \rho = 0$
$H_1: \rho \neq 0$

C. V. $= \pm 0.811$ d. f. $= 4$

Decision: Do not reject. There is not a significant linear relationship between class size and average grade.

29.

$$r = \frac{n(\sum xy)-(\sum x)(\sum y)}{\sqrt{[n(\sum x^2)-(\sum x)^2][n(\sum y^2)-(\sum y)^2]}}$$

$$r = \frac{5(180)-(15)(50)}{\sqrt{[5(55)-15^2][5(590)-50^2]}} = 1$$

$$r = \frac{5(180)-(50)(15)}{\sqrt{[5(590)-50^2][5(55)-15^2]}} = 1$$

The value of r does not change when the values for x and y are interchanged.

EXERCISE 10-2

1.
Draw the scatter plot and test the significance of the correlation coefficient.

3.
$y' = a + bx$

5.
It is the line that is drawn on the scatter plot such that the sum of the squares of the vertical distances each point is from the line is at a minimum.

7.
When r is positive, b will be positive.
When r is negative, b will be negative.

9.
The closer r is to $+1$ or -1, the more accurate the predicted value will be.

11.
$$a = \frac{(\sum y)(\sum x^2)-(\sum x)(\sum xy)}{n(\sum x^2)-(\sum x)^2}$$

$$a = \frac{(680.1)(142.52)-(31)(3202.71)}{8(142.52)-(31)^2} = -13.151$$

$$b = \frac{n(\sum xy)-(\sum x)(\sum y)}{n(\sum x^2)-(\sum x)^2}$$

$$b = \frac{8(3202.71)-(31)(680.1)}{8(142.52)-(31)^2} = 25.333$$

Chapter 10 - Correlation and Regression

11. continued

$y' = a + bx$

$y' = -13.151 + 25.333x$

$y' = -13.151 + 25.333(4.5) = 100.848$

robberies

13.

$a = \dfrac{(\sum y)(\sum x^2) - (\sum x)(\sum xy)}{n(\sum x^2) - (\sum x)^2}$

$a = \dfrac{(9283)(299,315) - (1045)(2,380,435)}{9(299,315) - (1045)^2} = 181.661$

$b = \dfrac{n(\sum xy) - (\sum x)(\sum y)}{n(\sum x^2) - (\sum x)^2}$

$b = \dfrac{9(2,380,435) - (1045)(9283)}{9(299,315) - (1045)^2} = 7.319$

$y' = a + bx$

$y' = 181.661 + 7.319x$

$y' = 181.661 + 7.319(200) = \$1645.5$

million gross receipts

15.

$a = \dfrac{(\sum y)(\sum x^2) - (\sum x)(\sum xy)}{n(\sum x^2) - (\sum x)^2}$

$a = \dfrac{(1105)(220) - (32)(3405)}{6(220) - (32)^2}$

$a = \dfrac{243100 - 108960}{1320 - 1024} = 453.176$

$b = \dfrac{n(\sum xy) - (\sum x)(\sum y)}{n(\sum x^2) - (\sum x)^2}$

$b = \dfrac{6(3405) - (32)(1105)}{6(220) - (32)^2}$

$b = -50.439$

$y' = a + bx$

$y' = 453.176 - 50.439x$

$y' = 453.176 - 50.439(4) = \$251.42$

17.

Since r is not significant, no regression should be done. $\bar{y}$ can be used for the prediction.

19.

$a = \dfrac{(\sum y)(\sum x^2) - (\sum x)(\sum xy)}{n(\sum x^2) - (\sum x)^2}$

$a = \dfrac{(383.5)(3430.87) - (130.9)(10,076.41)}{5(3430.87) - (130.9)^2}$

$a = -167.012$

$b = \dfrac{n(\sum xy) - (\sum x)(\sum y)}{n(\sum x^2) - (\sum x)^2}$

$b = \dfrac{5(10,076.41) - (130.9)(383.5)}{5(3430.87) - (130.9)^2}$

$b = 9.309$

$y' = a + bx$

$y' = -167.012 + 9.309x$

$y' = -167.012 + 9.309(26.8) = 82.5$

months

21.

For x = Students and y = Faculty:

$a = \dfrac{(\sum y)(\sum x^2) - (\sum x)(\sum xy)}{n(\sum x^2) - (\sum x)^2}$

$a = \dfrac{(970)(14,023,529) - (9689)(1,410,572)}{7(14,023,529) - (9689)^2}$

$a = -14.974$

$b = \dfrac{n(\sum xy) - (\sum x)(\sum y)}{n(\sum x^2) - (\sum x)^2}$

$b = \dfrac{7(1,410,572) - (9689)(970)}{7(14,023,529) - (9689)^2} = 0.111$

$y' = a + bx$

$y' = -14.974 + 0.111x$

23.

$a = \dfrac{(\sum y)(\sum x^2) - (\sum x)(\sum xy)}{n(\sum x^2) - (\sum x)^2}$

$a = \dfrac{(302.7)(25,192.41) - (367.5)(22,207.31)}{6(25,192.41) - (367.5)^2}$

$a = -33.261$

$b = \dfrac{n(\sum xy) - (\sum x)(\sum y)}{n(\sum x^2) - (\sum x)^2}$

$b = \dfrac{6(22,207.31) - (367.5)(302.7)}{6(25,192.41) - (367.5)^2}$

$b = 1.367$

$y\prime = a + bx$

$y\prime = -33.261 + 1.367x$

$y\prime = -33.261 + 1.367(80) = 76.099$ or 76.1%

25.
Since r is not significant, no regression should be done. $\bar{y}$ can be used for the prediction.

27.
Since r is not significant, no regression should be done. $\bar{y}$ can be used for the prediction.

29.

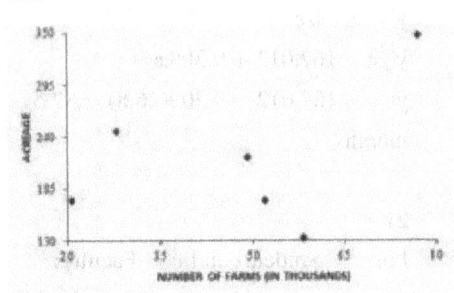

$\sum x = 285$
$\sum y = 1289$
$\sum x^2 = 15{,}637.88$
$\sum y^2 = 305{,}731$
$\sum xy = 64{,}565.8$
$n = 6$

$$r = \frac{n(\sum xy) - (\sum x)(\sum y)}{\sqrt{[n(\sum x^2) - (\sum x)^2][n(\sum y^2) - (\sum y)^2]}}$$

$$r = \frac{6(64{,}565.8) - (285)(1289)}{\sqrt{[6(15{,}637.88) - (285)^2][6(305{,}731) - (1289)^2]}}$$

$r = 0.429$

H_0: $\rho = 0$
H_1: $\rho \neq 0$
C. V. $= \pm 0.811$ d. f. $= 4$

Decision: Do not reject
There is no significant relationship between the number of farms and acreage.

31.

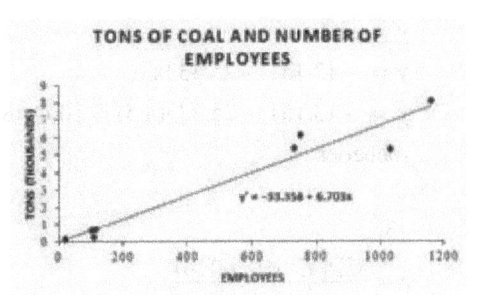

$\sum x = 4027$
$\sum y = 26{,}728$
$\sum x^2 = 3{,}550{,}103$
$\sum y^2 = 162{,}101{,}162$
$\sum xy = 23{,}663{,}669$
$n = 8$

$$r = \frac{n(\sum xy) - (\sum x)(\sum y)}{\sqrt{[n(\sum x^2) - (\sum x)^2][n(\sum y^2) - (\sum y)^2]}}$$

$$r = \frac{8(23662669) - (4027)(26728)}{\sqrt{[8(3550103) - 4027^2][8(162101162) - (26728)^2]}}$$

$r = 0.970$

H_0: $\rho = 0$
H_1: $\rho \neq 0$
C. V. $= \pm 0.707$ d. f. $= 6$

Decision: Reject. There is a significant relationship between the number of tons of coal produced and the number of employees.

$$a = \frac{(\sum y)(\sum x^2) - (\sum x)(\sum xy)}{n(\sum x^2) - (\sum x)^2}$$

$$a = \frac{(26728)(3550103) - (4027)(23663669)}{8(3550103) - (4027)^2}$$

$a = -33.358$

$$b = \frac{n(\sum xy) - (\sum x)(\sum y)}{n(\sum x^2) - (\sum x)^2}$$

$$b = \frac{8(23663669) - (4027)(26728)}{8(3550103) - (4027)^2} = 6.703$$

$y' = a + bx$
$y' = -33.358 + 6.703x$
$y' = -33.358 + 6.703(500) = 3318.142$

33.

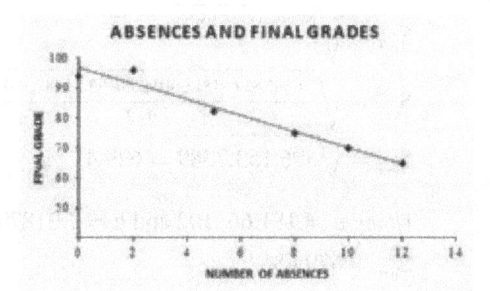

ABSENCES AND FINAL GRADES (scatter plot)

$\sum x = 37$
$\sum y = 482$
$\sum x^2 = 337$
$\sum y^2 = 39526$
$\sum xy = 2682$
n = 6

$$r = \frac{n(\sum xy) - (\sum x)(\sum y)}{\sqrt{[n(\sum x^2) - (\sum x)^2][n(\sum y^2) - (\sum y)^2]}}$$

$$r = \frac{6(2682) - (37)(482)}{\sqrt{[6(337) - (37)^2][6(39526) - (482)^2]}}$$

$r = -0.981$

H_0: $\rho = 0$
H_1: $\rho \neq 0$
C. V. = ± 0.811 d. f. = 4

Decision: Reject. There is a significant relationship between the number of absences and the final grade.

$$a = \frac{(\sum y)(\sum x^2) - (\sum x)(\sum xy)}{n(\sum x^2) - (\sum x)^2}$$

$$a = \frac{(482)(337) - (37)(2682)}{6(337) - (37)^2} = 96.784$$

$$b = \frac{n(\sum xy) - (\sum x)(\sum y)}{n(\sum x^2) - (\sum x)^2}$$

$$b = \frac{6(2682) - (37)(482)}{6(337) - (37)^2} = -2.668$$

$y' = a + bx$
$y' = 96.784 - 2.668x$

35.

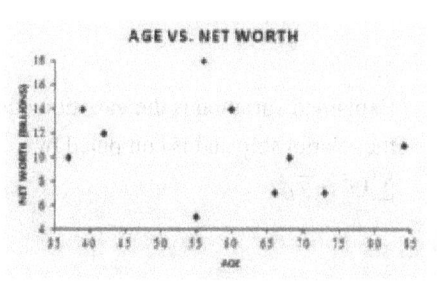

AGE VS. NET WORTH (scatter plot)

$\sum x = 580$
$\sum y = 108$
$\sum x^2 = 35,780$
$\sum y^2 = 1304$
$\sum xy = 6120$
n = 10

$$r = \frac{n(\sum xy) - (\sum x)(\sum y)}{\sqrt{[n(\sum x^2) - (\sum x)^2][n(\sum y^2) - (\sum y)^2]}}$$

$$r = \frac{10(6120) - (580)(108)}{\sqrt{[10(35,780) - 580^2][10(1304) - 108^2]}}$$

$r = -0.265$

H_0: $\rho = 0$
H_1: $\rho \neq 0$

$t = -0.777$; P-value > 0.05 (0.459)
Decision: Do not reject since P-value > 0.05. There is no significant linear relationship between the ages of billionaires and their net worth. Since r is not significant, no regression should be done.

37.
For Exercise 15:
$\bar{x} = 5.3333$
$\bar{y} = 184.1667$
$b = -50.439$
$a = \bar{y} - b\bar{x}$
$a = 184.1667 - (-50.439)(5.3333)$
$a = 184.1667 + 269.0063$
$a = 453.173$ (differs due to rounding)

For Exercise 16:
Since r is not significant, no regression should be done.

Chapter 10 - Correlation and Regression

EXERCISE SET 10-3

1.
Explained variation is the variation due to the relationship and is computed by $\sum(y' - \bar{y})^2$.

3.
Total variation is the sum of the squares of the vertical distances of the points from the mean. It is computed by $\sum(y - \bar{y})^2$.

5.
It is found by squaring r.

7.
The coefficient of non-determination is $1 - r^2$.

9.
For $r = 0.44$, $r^2 = 0.1936$ and $1 - r^2 = 0.8064$.
Thus 19.36% of the variation of y is due to the variation of x, and 80.64% of the variation of y is due to chance.

11.
For $r = 0.97$, $r^2 = 0.9409$ and $1 - r^2 = 0.0591$. Thus 94.09% of the variation of y is due to the variation of x, and 5.91% of the variation of y is due to chance.

13.
For $r = 0.15$, $r^2 = 0.0225$ and $1 - r^2 = 0.9775$. Thus 2.25% of the variation of y is due to the variation of x, and 97.75% of the variation of y is due to chance.

15.
$$S_{est} = \sqrt{\frac{\sum y^2 - a\sum y - b\sum xy}{n-2}}$$

$$S_{est} = \sqrt{\frac{21{,}881{,}839 - 181.661(9283) - (7.319)(2{,}380{,}435)}{9-2}}$$

$$S_{est} = \sqrt{396{,}153.7389} = 629.41$$

Using $a = 181.661102$ and $b = 7.318708$,
$S_{est} = 629.4862$

17.
$$S_{est} = \sqrt{\frac{\sum y^2 - a\sum y - b\sum xy}{n-2}}$$

$$S_{est} = \sqrt{\frac{364525 - (453.176)(1105) - (-50.439)(3405)}{6-2}}$$

$$S_{est} = 94.22$$

19.
$y' = 181.661 + 7.319x$
$y' = 181.661 + 7.319(200)$
$y' = 1645.461$

$$y' - t_{\frac{\alpha}{2}} \cdot S_{est}\sqrt{1 + \frac{1}{n} + \frac{n(x - \bar{X})}{n\sum x^2 - (\sum x)^2}} < y <$$

$$y' + t_{\frac{\alpha}{2}} \cdot S_{est}\sqrt{1 + \frac{1}{n} + \frac{n(x - \bar{X})^2}{n\sum x^2 - (\sum x)^2}}$$

$1645.461 - (1.895)(629.4862)$
$\sqrt{1 + \frac{1}{9} + \frac{9(200 - 116.11)^2}{9(299{,}315) - 1045^2}}$
$< y < 1645.461 +$
$(1.895)(629.4862)\sqrt{1 + \frac{1}{9} + \frac{9(200 - 116.11)^2}{9(299{,}315) - 1045^2}}$

$1645.461 - 1279.580227 < y <$
$1645.461 + 1279.580227$

$365.88 < y < 2925.04$

Chapter 10 - Correlation and Regression

21.

$y' = 453.176 - 50.439x$

$y' = 453.176 - 50.439(4)$

$y' = 251.42$

$y' - t_{\frac{\alpha}{2}} \cdot s_{est}\sqrt{1 + \frac{1}{n} + \frac{n(x-\overline{X})^2}{n\Sigma x^2 - (\Sigma x)^2}} < y <$

$\qquad y' + t_{\frac{\alpha}{2}} \cdot s_{est}\sqrt{1 + \frac{1}{n} + \frac{n(x-\overline{X})^2}{n\Sigma x^2 - (\Sigma x)^2}}$

$251.42 - 2.132(94.22)\sqrt{1 + \frac{1}{6} + \frac{6(4-5.33)^2}{6(220)-32^2}}$

$< y < 251.42 + 2.132(94.22)\sqrt{1 + \frac{1}{6} + \frac{6(4-5.33)^2}{6(220)-32^2}}$

$251.42 - (2.132)(94.22)(1.1) < y <$
$\qquad 251.42 + (2.132)(94.22)(1.1)$

$\$30.46 < y < \$472.38$

EXERCISE SET 10-4

1.
Simple linear regression has one independent variable and one dependent variable. Multiple regression has one dependent variable and two or more independent variables.

3.
The relationship would include all variables in one equations.

5.
The multiple correlation coefficient R is always higher than the individual correlation coefficients. Also, the value of R can range from 0 to +1.

7.

$y' = 0.217 + 0.0654x_1 + 0.32x_2$

$y' = 0.217 + 0.0654(72) + 0.32x_2(8) = 7.5$

9.

$y' = -14.9 + 0.93359x_1 + 0.99847x_2$
$\qquad + 5.3844x_3$

$y' = -14.9 + 0.93359(8) + 0.99847(34)$
$\qquad + 5.3844(11)$

$y' = 85.75$ (grade) or 86

11.
R is a measure of the strength of the relationship between the dependent variables and all the independent variables.

13.
R^2 is the coefficient of multiple determination. R^2_{adj} is adjusted for sample size and the number of predictors.

15.
The F test is used to test the significance of R.

REVIEW EXERCISES - CHAPTER 10

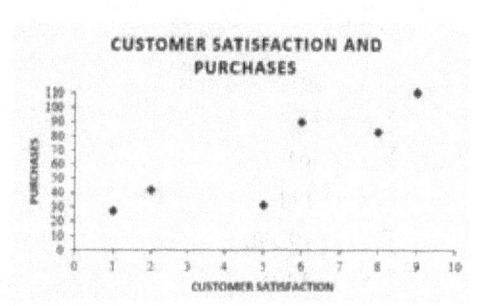

1.

CUSTOMER SATISFACTION AND PURCHASES

$\Sigma x = 31$

$\Sigma y = 383$

$\Sigma x^2 = 211$

$\Sigma y^2 = 30{,}543$

$\Sigma xy = 2460$

$n = 6$

$r = \dfrac{n(\Sigma xy) - (\Sigma x)(\Sigma y)}{\sqrt{[n(\Sigma x^2) - (\Sigma x)^2][n(\Sigma y^2) - (\Sigma y)^2]}}$

Chapter 10 - Correlation and Regression

1. continued

$$r = \frac{6(2460)-(31)(383)}{\sqrt{[6(211)-(31)^2][6(30{,}543)-(383)^2]}}$$

$r = 0.864$

H_0: $\rho = 0$

H_1: $\rho \neq 0$

C. V. $= \pm 0.917$ at $\alpha = 0.01$ d. f. $= 4$

Decision: Do not reject. There is no significant linear relationship between customer satisfaction and the amount customers spend. No regression should be done since r is not significant.

3.

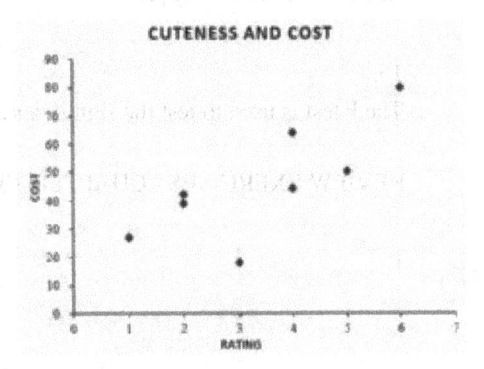

CUTENESS AND COST

$\sum x = 27$
$\sum y = 364$
$\sum x^2 = 111$
$\sum y^2 = 19{,}270$
$\sum xy = 1405$
$n = 8$

$$r = \frac{n(\sum xy)-(\sum x)(\sum y)}{\sqrt{[n(\sum x^2)-(\sum x)^2][n(\sum y^2)-(\sum y)^2]}}$$

$$r = \frac{8(1405)-(27)(364)}{\sqrt{[8(111)-(27)^2][8(19{,}270)-(364)^2]}}$$

$r = 0.761$

H_0: $\rho = 0$

H_1: $\rho \neq 0$

C. V. $= \pm 0.834$ d. f. $= 6$

3. continued

Decision: Do not reject. There is no significant linear relationship between the cuteness of a puppy and its cost. No regression should be done since r is not significant.

5.

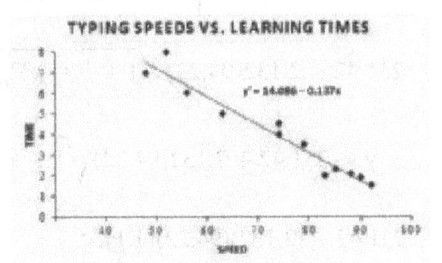

TYPING SPEEDS VS. LEARNING TIMES

$\sum x = 884$
$\sum y = 47.8$
$\sum x^2 = 67{,}728$
$\sum y^2 = 242.06$
$\sum xy = 3163.8$
$n = 12$

$$r = \frac{n(\sum xy)-(\sum x)(\sum y)}{\sqrt{[n(\sum x^2)-(\sum x)^2][n(\sum y^2)-(\sum y)^2]}}$$

$$r = \frac{12(3163.8)-(884)(47.8)}{\sqrt{[12(67728)-(884)^2][12(242.06)-(47.8)^2]}}$$

$r = -0.974$

H_0: $\rho = 0$

H_1: $\rho \neq 0$

C. V. $= \pm 0.708$ d. f. $= 10$

Decision: Reject. There is a significant relationship between speed and time.

$$a = \frac{(\sum y)(\sum x^2)-(\sum x)(\sum xy)}{n(\sum x^2)-(\sum x)^2}$$

$$a = \frac{(47.8)(67728)-(884)(3163.8)}{12(67728)-(884)^2}$$

$a = 14.086$

$$b = \frac{n(\sum xy)-(\sum x)(\sum y)}{n(\sum x^2)-(\sum x)^2}$$

$$b = \frac{12(3163.8)-(884)(47.8)}{12(67728)-(884)^2}$$

Chapter 10 - Correlation and Regression

5. continued

$b = -0.137$

$y' = a + bx$

$y' = 14.086 - 0.137x$

$y' = 14.086 - 0.137(72) = 4.2$ hours

7.

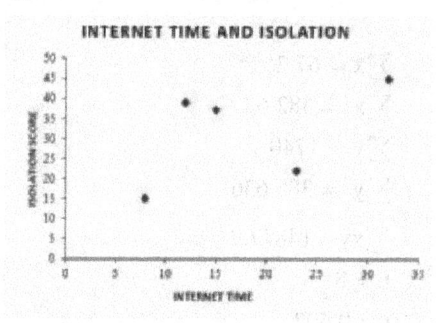

INTERNET TIME AND ISOLATION

$\sum x = 90$

$\sum y = 158$

$\sum x^2 = 1986$

$\sum y^2 = 5624$

$\sum xy = 3089$

$n = 5$

$r = \dfrac{n(\sum xy) - (\sum x)(\sum y)}{\sqrt{[n(\sum x^2) - (\sum x)^2][n(\sum y^2) - (\sum y)^2]}}$

$r = \dfrac{5(3089) - (90)(158)}{\sqrt{[5(1986) - (90)^2][5(5624) - (158)^2]}}$

$r = 0.510$

$H_0: \rho = 0$

$H_1: \rho \neq 0$

C. V. $= \pm 0.959$ d. f. $= 3$

Decision: Do not reject. There is no significant linear relationship between internet use and isolation. No regression should be done since r is not significant.

9.

$S_{est} = \sqrt{\dfrac{\sum y^2 - a\sum y - b\sum xy}{n-2}}$

$S_{est} = \sqrt{\dfrac{242.06 - 14.086(47.8) + 0.137(3163.8)}{12 - 2}}$

9. continued

$S_{est} = \sqrt{\dfrac{2.1898}{10}} = \sqrt{0.21898} = 0.468$

(Note: TI-83 calculator answer is 0.513)

11.

(For calculation purposes only, since no regression should be done.)

$y' = 14.086 - 0.137x$

$y' = 14.086 - 0.137(72) = 4.222$

$y' - t_{\frac{\alpha}{2}} \cdot S_{est}\sqrt{1 + \dfrac{1}{n} + \dfrac{n(x-\overline{X})^2}{n\Sigma x^2 - (\Sigma x)^2}} < y <$

$y' + t_{\frac{\alpha}{2}} \cdot S_{est}\sqrt{1 + \dfrac{1}{n} + \dfrac{n(x-\overline{X})^2}{n\Sigma x^2 - (\Sigma x)^2}}$

$4.222 - 1.812(0.468)\sqrt{1 + \dfrac{1}{12} + \dfrac{12(72-73.667)^2}{12(67,728) - 884^2}}$

$< y < 4.222 + 1.812(0.468)\sqrt{1 + \dfrac{1}{12} + \dfrac{12(72-73.667)^2}{12(67,728) - 884^2}}$

$4.222 - 1.812(0.468)(1.041) < y <$
$\qquad 4.222 + 1.812(0.468)(1.041)$

$3.34 < y < 5.10$

13.

$y' = 12.8 + 2.09X_1 + 0.423X_2$

$y' = 12.8 + 2.09(4) + 0.423(2) = 22.006$ or 22.01

15.

$R^2_{adj} = 1 - \left[\dfrac{(1-R^2)(n-1)}{n-k-1}\right]$

$R^2_{adj} = 1 - \left[\dfrac{(1-0.873^2)(10-1)}{10-3-1}\right]$

$R^2_{adj} = 1 - \left[\dfrac{2.1408}{6}\right] = 0.643$

CHAPTER 10 QUIZ

1. False, the y variable would decrease.

3. True

5. False, a relationship may be caused by chance.

7. a

9. d

11. b

13. Independent

15. b (slope)

17. + 1, − 1

19.

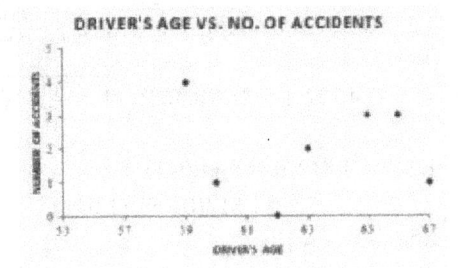

DRIVER'S AGE VS. NO. OF ACCIDENTS

$\sum x = 442$
$\sum x^2 = 27,964$
$\sum y = 14$
$\sum y^2 = 40$
$\sum xy = 882$
n = 7
r = − 0.078
H₀: $\rho = 0$
H₁: $\rho \neq 0$
C. V. = ± 0.754   d. f. = 5
Decision: Do not reject. There is not a significant relationship between age and number of accidents. No regression should be done.

21.

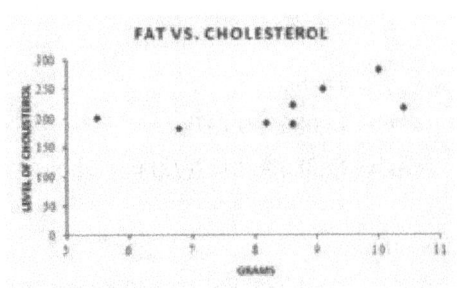

FAT VS. CHOLESTEROL

$\sum x = 67.2$
$\sum x^2 = 582.62$
$\sum y = 1740$
$\sum y^2 = 386,636$
$\sum xy = 14847.9$
n = 8
r = 0.602
H₀: $\rho = 0$
H₁: $\rho \neq 0$
C. V. = ± 0.707   d. f. = 6
Decision: Do not reject. There is no significant linear relationship between fat and cholesterol. No regression should be done.

23.
(For calculation purposes only, since no regression should be done.)

$$S_{est} = \sqrt{\frac{386,636 - 110.12(1740) - 12.784(14,847.9)}{8-2}}$$

$S_{est} = 29.47^*$

25.
Since no regression should be done, the average of the y′ values is used: $\bar{y} = 217.5$

27.

$$R = \sqrt{\frac{(0.561)^2 + (0.714)^2 - 2(0.561)(0.714)(0.625)}{1 - (0.625)^2}}$$

$R = 0.729^*$

*These answers may vary due to method of calculation and/or rounding.

Chapter 11 - Other Chi-Square Tests

Note: Graphs are not to scale and are intended to convey a general idea.

Answers may vary due to rounding, TI-83's, or computer programs.

EXERCISE SET 11-1

1.
The variance test compares a sample variance to a hypothesized population variance; the goodness-of-fit test compares a distribution obtained from a sample with a hypothesized distribution.

3.
The expected values are computed based on what the null hypothesis states about the distribution.

5.
H_0: The students show no preference for class times.
H_1: The students show a preference for class times. (claim)

C. V. = 11.345   d. f. = 3 $\alpha = 0.01$

$E = \frac{116}{4} = 29$

$\chi^2 = \frac{(24-29)^2}{29} + \frac{(35-29)^2}{29} + \frac{(31-29)^2}{29} + \frac{(26-29)^2}{29}$

$\chi^2 = 2.552$

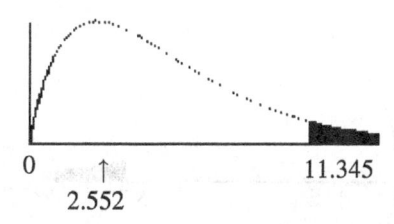

Do not reject the null hypothesis. There is not enough evidence to support the claim that that the students show a preference for class times.

7.
H_0: The distribution is as follows: 45% favor extending the school year, 47% do not want the school year extended, and 8% have no opinion.
H_1: The distribution is not the same as that stated in the null hypothesis. (claim)

C. V. = 5.991   d. f. = 2 $\alpha = 0.05$

O	E
46	0.45(100) = 45
42	0.47(100) = 47
12	0.08(100) = 8

$\chi^2 = \sum \frac{(O-E)^2}{E} = \frac{(46-45)^2}{45} + \frac{(42-47)^2}{47} + \frac{(12-8)^2}{8} = 2.554$

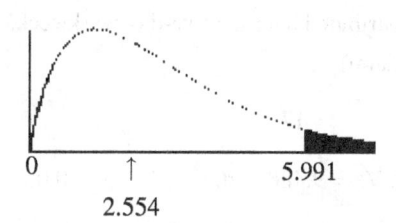

Do not reject the null hypothesis. There is not enough evidence to support the claim that the percentages are different from the ones stated in the null hypothesis.

9.
H_0: The proportions are distributed as follows: safe, 35%; not safe, 52%; no opinion, 13%.
H_1: The distribution is not the same as stated in the null hypothesis. (claim)

C. V. = 9.210   d. f. = 2 $\alpha = 0.01$

O	E
40	0.35(120) = 42
60	0.52(120) = 62.4
20	0.13(120) = 15.6

$\chi^2 = \frac{(40-42)^2}{42} + \frac{(60-62.4)^2}{62.4} + \frac{(20-15.6)^2}{15.6}$

$\chi^2 = 1.429$

Chapter 11 - Other Chi-Square Tests

9. continued

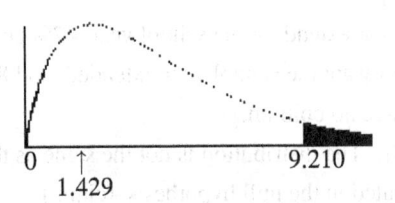

0 ↑ 9.210
1.429

Do not reject the null hypothesis. There is not enough evidence to support the claim that the proportions are different.

11.
H_0: Employee absences are equally distributed over the five-day workweek.
H_1: Employee absences are not equally distributed over the five-day workweek. (claim)

$E = \frac{85}{5} = 17$

C. V. = 9.488   d. f. = 4 $\alpha = 0.05$

$\chi^2 = \frac{(13-17)^2}{17} + \frac{(10-17)^2}{17} + \frac{(16-17)^2}{17} + \frac{(22-17)^2}{17} + \frac{(24-17)^2}{17}$

$\chi^2 = 8.235$

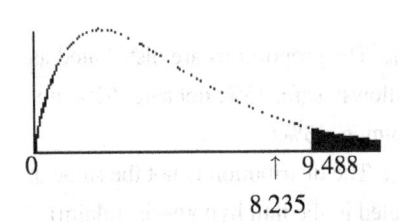

0 ↑ 9.488
8.235

Do not reject the null hypothesis. There is not enough evidence to say that the absences are not equally distributed during the week.

13.
H_0: 10% of deaths were ages 0 - 19, 50% were ages 20 - 44, and 40% were ages 45 and older.
H_1: The distribution is not the same as stated in the null hypothesis. (claim)

13. continued

O	E
13	0.10(100) = 10
62	0.50(100) = 50
25	0.40(100) = 40

C. V. = 5.991   d. f. = 2 $\alpha = 0.05$

$\chi^2 = \frac{(13-10)^2}{10} + \frac{(62-50)^2}{50} + \frac{(25-40)^2}{40}$

$\chi^2 = 9.405$

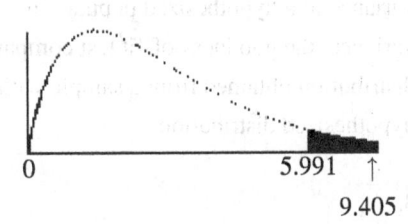

0   5.991 ↑
           9.405

Reject the null hypothesis. There is enough evidence to support the claim that the proportions are different from those stated by the National Safety Council.

15.
H_0: The proportion of Internet users is the same for all groups.
H_1: The proportion of Internet users is not the same for all groups. (claim)

$E = \frac{125}{3} = 41.67$

C. V. = 5.991   d. f. = 2 $\alpha = 0.05$

$\chi^2 = \sum \frac{(O-E)^2}{E} = \frac{(44-41.67)^2}{41.67} + \frac{(41-41.67)^2}{41.67} + \frac{(40-41.67)^2}{41.67}$

$= 0.208$

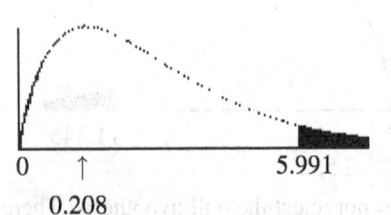

0 ↑   5.991
0.208

Do not reject the null hypothesis. There is not enough evidence to support the claim that the proportions differ.

Chapter 11 - Other Chi-Square Tests

17.

H_0: The distribution of the ways people pay for their prescriptions is as follows: 60% used personal funds, 25% used insurance, and 15% used Medicare. (claim)

H_1: The distribution is not the same as stated in the null hypothesis.

$\alpha = 0.05$ d.f. = 2
P-value > 0.05
(TI: P-value = 0.7164)

O	E
32	0.6(50) = 30
10	0.25(50) = 12.5
8	0.15(50) = 7.5

$$\chi^2 = \sum \frac{(O-E)^2}{E} = \frac{(32-30)^2}{30} + \frac{(10-12.5)^2}{12.5}$$
$$+ \frac{(8-7.5)^2}{7.5} = 0.667$$

Do not reject the null hypothesis since P-value > 0.05. There is not enough evidence to reject the claim that the distribution is the same as stated in the null hypothesis. An implication of the results is that the majority of people are using their own money to pay for medications. A less expensive medication could help people financially.

19.

H_0: The coins are balanced and randomly tossed. (claim)

H_1: The distribution is not the same as stated in the null hypothesis.

C.V. = 7.815   d.f. = 3

E(0) = 0.125(72) = 9
E(1) = 0.375(72) = 27
E(2) = 0.375(72) = 27
E(3) = 0.125(72) = 9
(use the binomial distribution with n = 3 and p = 0.05)

$$\chi^2 = \frac{(3-9)^2}{9} + \frac{(10-27)^2}{27} + \frac{(17-27)^2}{27}$$
$$+ \frac{(42-9)^2}{9} = 139.4$$

19. continued

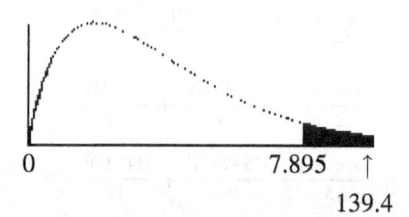

0   7.895 ↑
       139.4

Reject the null hypothesis. There is enough evidence to reject the claim that the coins are balanced and randomly tossed.

EXERCISE SET 11-2

1.

The independence test and the goodness of fit test both use the same formula for computing the test-value; however, the independence test uses a contingency table whereas the goodness of fit test does not.

3.

H_0: The variables are independent or not related.

H_1: The variables are dependent or related.

5.

The expected values are computed as (row total · column total) ÷ grand total.

7.

H_0: The living arrangement of a person is independent of the gender of the person.

H_1: The living arrangement of a person is dependent upon the gender of the person. (claim)

C.V. = 7.815   d.f. = 3 $\alpha = 0.05$

	Spouse	Relative	Nonrelative	Alone
Men	57(55)	8(6.5)	25(26.5)	10(12)
Women	53(55)	5(6.5)	28(26.5)	14(12)

7. continued

$$\chi^2 = \sum \frac{(O-E)^2}{E} = \frac{(57-55)^2}{55} + \frac{(8-6.5)^2}{6.5}$$

$$+ \frac{(25-26.5)^2}{26.5} + \frac{(10-12)^2}{12} + \frac{(53-55)^2}{55}$$

$$+ \frac{(5-6.5)^2}{6.5} + \frac{(28-26.5)^2}{26.5} + \frac{(14-12)^2}{12}$$

$$\chi^2 = 1.674$$

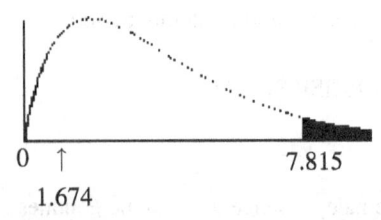

Critical value 7.815; test statistic 1.674.

Do not reject the null hypothesis. There is not enough evidence to support the claim that the living arrangement of a person is dependent on the gender of the individual.

9.

H_0: Pet ownership is independent of the number of persons living in the household.

H_1: Pet ownership is dependent on the number of persons living in the household. (claim)

C. V. = 6.251   d. f. = 3 $\alpha = 0.10$

	1 person	2 people	3 people	4 or more people
Dog	7(8)	16(15)	11(13.5)	16(13.5)
Cat	9(8)	14(15)	16(13.5)	11(13.5)

$$\chi^2 = \sum \frac{(O-E)^2}{E} = \frac{(7-8)^2}{8} + \frac{(16-15)^2}{15}$$

$$+ \frac{(11-13.5)^2}{13.5} + \frac{(16-13.5)^2}{13.5} + \frac{(9-8)^2}{8}$$

$$+ \frac{(14-15)^2}{15} + \frac{(16-13.5)^2}{13.5} + \frac{(11-13.5)^2}{13.5}$$

$$\chi^2 = 2.235$$

9. continued

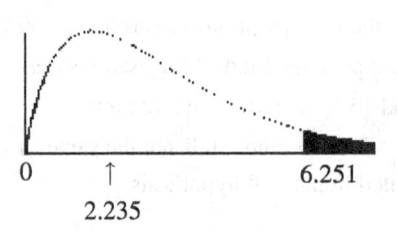

Critical value 6.251; test statistic 2.235.

Do not reject the null hypothesis. There is not enough evidence to support the claim that pet ownership is dependent on the number of persons living in the household.

11.

H_0: The types of violent crimes committed are independent of the cities where they are committed.

H_1: The The types of violent crimes committed are dependent upon the cities where they are committed. (claim)

C. V. = 12.592   d. f. = 6 $\alpha = 0.05$

$$E = \frac{(\text{row sum})(\text{column sum})}{\text{grand total}}$$

$$E_{1,1} = \frac{(119)(85)}{502} = 20.1494$$

$$E_{1,2} = \frac{(119)(141)}{502} = 33.4243$$

$$E_{1,3} = \frac{(119)(276)}{502} = 65.4263$$

$$E_{2,1} = \frac{(119)(85)}{502} = 20.1494$$

$$E_{2,2} = \frac{(119)(141)}{502} = 33.4243$$

$$E_{2,3} = \frac{(119)(276)}{502} = 65.4263$$

$$E_{3,1} = \frac{(128)(85)}{502} = 21.6733$$

$$E_{3,2} = \frac{(128)(141)}{502} = 35.9522$$

$$E_{3,3} = \frac{(128)(276)}{502} = 70.3745$$

$$E_{4,1} = \frac{(136)(85)}{502} = 23.0279$$

Chapter 11 - Other Chi-Square Tests

11. continued

$$E_{4,2} = \frac{(136)(141)}{502} = 38.1992$$

$$E_{4,3} = \frac{(136)(276)}{502} = 74.7729$$

City	Rape	Robbery	Assault
Cary, NC	14(20.1494)	35(33.4243)	70(65.4263)
Amherst, NY	10(20.1494)	33(33.4243)	76(65.4263)
Simi Valley, CA	14(21.6733)	37(35.9522)	77(70.3745)
Norman, OK	47(23.0279)	36(38.1992)	53(74.7729)

$$\chi^2 = \sum \frac{(O-E)^2}{E} = \frac{(14-20.1494)^2}{20.1494}$$
$$+ \frac{(35-33.4243)^2}{33.4243} + \frac{(70-65.4263)^2}{65.4263}$$
$$+ \frac{(10-20.1494)^2}{20.1494} + \frac{(33-33.4243)^2}{33.4243}$$
$$+ \frac{(76-65.4263)^2}{65.4263} + \frac{(14-21.6733)^2}{21.6733}$$
$$+ \frac{(37-35.9522)^2}{35.9522} + \frac{(77-70.3745)^2}{70.3745}$$
$$+ \frac{(47-23.0279)^2}{23.0279} + \frac{(36-38.1992)^2}{38.1992}$$
$$+ \frac{(53-74.7729)^2}{74.7729}$$

$$\chi^2 = 43.890$$

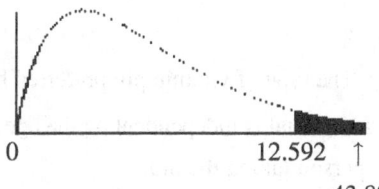

Reject the null hypothesis. There is enough evidence to support the claim that the types of violent crimes are dependent upon the cities where they are committed.

13.
H_0: The length of unemployment time is independent of the type of industry where the worker is employed.
H_1: The length of unemployment time is dependent upon the type of industry where the worker is employed. (claim)

C. V. = 9.488   d. f. = 4 $\alpha = 0.05$

13. continued

	< 5 wks	5 - 14 wks	15 - 26 wks
Trans./Util.	85(81.0368)	110(104.2974)	80(89.6658)
Information	48(44.2019)	57(56.8895)	45(48.9086)
Financial	83(90.7613)	111(116.8131)	114(100.4256)

$$\chi^2 = \sum \frac{(O-E)^2}{E} = \frac{(85-81.0368)^2}{81.0368}$$
$$+ \frac{(110-104.2974)^2}{104.2974} + \frac{(80-89.6658)^2}{89.6658}$$
$$+ \frac{(48-44.2019)^2}{44.2019} + \frac{(57-56.8895)^2}{56.8895}$$
$$+ \frac{(45-48.9086)^2}{48.9086} + \frac{(83-90.7613)^2}{90.7613}$$
$$+ \frac{(111-116.8131)^2}{116.8131} + \frac{(114-100.4256)^2}{100.4256}$$

$$\chi^2 = 4.974$$

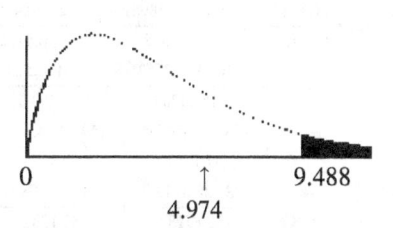

Do not reject the null hypothesis. There is not enough evidence to support the claim that the length of unemployment time is dependent upon the type of industry where the worker is employed.

15.
H_0: The program of study of a student is independent of the type of institution.
H_1: The program of study of a student is dependent upon the type of institution. (claim)

C. V. = 7.815   d. f. = 3 $\alpha = 0.05$

$$E_{1,1} = \frac{(88)(302)}{707} = 37.5898$$

$$E_{1,2} = \frac{(88)(405)}{707} = 50.4102$$

$$E_{2,1} = \frac{(441)(302)}{707} = 188.3762$$

$$E_{2,2} = \frac{(441)(405)}{707} = 252.6238$$

15. continued

$$E_{3,1} = \frac{(87)(302)}{707} = 37.1627$$

$$E_{3,2} = \frac{(87)(405)}{707} = 49.8373$$

$$E_{4,1} = \frac{(91)(302)}{707} = 38.8713$$

$$E_{4,2} = \frac{(91)(405)}{707} = 52.1287$$

	Two-year	Four-year
Agriculture	36(37.5898)	52(50.4102)
Criminal Justice	210(188.3762)	231(252.6238)
Lang/Lit	28(37.1627)	59(49.8373)
Math/Stat	28(38.8713)	63(52.1287)

$$\chi^2 = \sum \frac{(O-E)^2}{E} = \frac{(36-37.5898)^2}{37.5898} + \frac{(52-50.4102)^2}{50.4102}$$
$$+ \frac{(210-188.3762)^2}{188.3762} + \frac{(231-252.6238)^2}{252.6238}$$
$$+ \frac{(28-37.1627)^2}{37.1627} + \frac{(59-49.8373)^2}{49.8373}$$
$$+ \frac{(28-38.8713)^2}{38.8713} + \frac{(63-52.1287)^2}{52.1287}$$

$$\chi^2 = 13.702$$

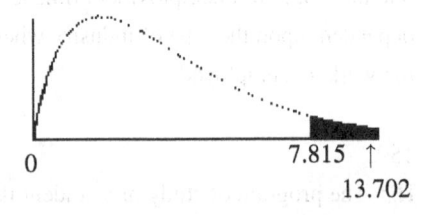

Reject the null hypothesis. There is enough evidence to conclude that the type of program is dependent on the type of institution.

17.

H_0: The type of automobile owned by a person is independent of the gender of the individual.

H_1: The type of automobile owned by a person is dependent on the gender of the individual. (claim)

C. V. = 6.251   d. f. = 3 $\alpha = 0.10$

17. continued

	Luxury	Large	Midsize	Small
Men	15(12.5654)	9(7.8534)	49(58.1152)	27(21.4660)
Women	9(11.4346)	6(7.1466)	62(52.8848)	14(19.5340)

$$\chi^2 = \frac{(15-12.5654)^2}{12.5654} + \frac{(9-7.8534)^2}{7.8534}$$
$$+ \frac{(49-58.1152)^2}{58.1152} + \frac{(27-21.4660)^2}{21.4660}$$
$$+ \frac{(9-11.4346)^2}{11.4346} + \frac{(6-7.1466)^2}{7.1466}$$
$$+ \frac{(62-52.8848)^2}{52.8848} + \frac{(14-19.5340)^2}{19.5340}$$

$$\chi^2 = 7.337$$

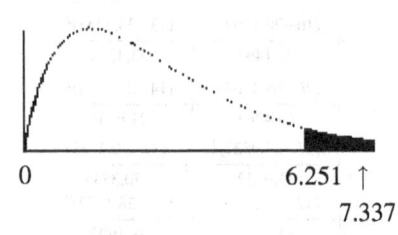

Reject the null hypothesis. There is enough evidence to support the claim that the type of automobile is related to the gender of the owner.

19.

H_0: The type of vitamin pill preferred by an individual is independent on the age of the person taking the pill.

H_1: The type of vitamin pill preferred by the individual is dependent on the age of the individual. (claim)

C. V. = 7.779 $\alpha = 0.10$ d. f. = 4

	Liquid	Tablet	Gummy
20–39	5(8.827)	16(12.202)	6(5.971)
40–59	10(15.365)	25(21.24)	12(10.394)
60–over	19(9.808)	6(13.558)	5(6.635)

$$\chi^2 = \sum \frac{(O-E)^2}{E} = \frac{(5-8.827)^2}{8.827} + \frac{(16-12.202)^2}{12.202}$$
$$+ \frac{(6-5.971)^2}{5.971} + \frac{(10-15.365)^2}{15.365} + \frac{(25-21.24)^2}{21.24}$$
$$+ \frac{(12-10.394)^2}{10.394} + \frac{(19-9.808)^2}{9.808} + \frac{(6-13.558)^2}{13.558}$$
$$+ \frac{(5-6.635)^2}{6.635}$$

Chapter 11 - Other Chi-Square Tests

19. continued

$\chi^2 = 18.859$

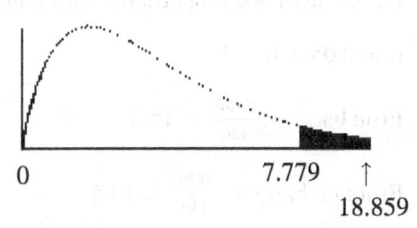

Reject the null hypothesis. There is enough evidence to support the claim that the type of vitamin pill preferred is dependent upon the age of the individual.

21.

H_0: $p_1 = p_2 = p_3$ (claim)

H_1: At least one proportion is different.

C. V. = 4.605   d. f. = 2

$E(\text{Cesarean}) = \frac{100(111)}{300} = 37$

$E(\text{Non-cesarean}) = \frac{100(189)}{300} = 63$

	Hosp. A	Hosp. B	Hosp. C
Cesarean	44(37)	28(37)	39(37)
Non-Cesarean	56(63)	72(63)	61(63)

$\chi^2 = \frac{(44-37)^2}{37} + \frac{(28-37)^2}{37} + \frac{(39-37)^2}{37}$
$+ \frac{(56-63)^2}{63} + \frac{(72-63)^2}{63} + \frac{(61-63)^2}{63}$

$\chi^2 = 5.749$

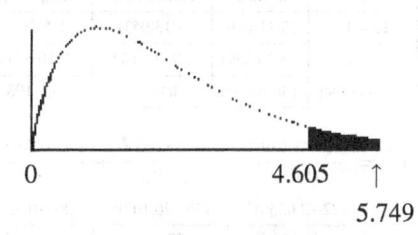

Reject the null hypothesis. There is enough evidence to reject the claim that the proportions are equal.

23.

H_0: $p_1 = p_2 = p_3 = p_4$ (claim)

H_1: At least one proportion is different.

C. V. = 7.815   d. f. = 3

$E(\text{passed}) = \frac{120(167)}{120} = 41.75$

$E(\text{failed}) = \frac{120(313)}{120} = 78.25$

	Southside	West End	East Hills	Jefferson
Passed	49(41.75)	38(41.75)	46(41.75)	34(41.75)
Failed	71(78.25)	82(78.25)	74(78.25)	86(78.25)

$\chi^2 = \frac{(49-41.75)^2}{41.75} + \frac{(38-41.75)^2}{41.75} + \frac{(46-41.75)^2}{41.75}$
$+ \frac{(34-41.75)^2}{41.75} + \frac{(71-78.25)^2}{78.25} + \frac{(82-78.25)^2}{78.25}$
$+ \frac{(74-78.25)^2}{78.25} + \frac{(86-78.25)^2}{78.25}$

$\chi^2 = 5.317$

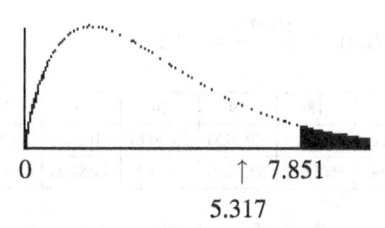

Do not reject the null hypothesis. There is not enough evidence to reject the claim that the proportions are equal.

25.

H_0: $p_1 = p_2 = p_3$ (claim)

H_1: At least one proportion is different from the others.

C. V. = 5.991   d. f. = 2

	Pork	Beef	Poultry
Athletes	15(16)	36(32)	9(12)
Nonathletes	17(16)	28(32)	15(12)

$\chi^2 = \frac{(15-16)^2}{16} + \frac{(36-32)^2}{32} + \frac{(9-12)^2}{12}$
$+ \frac{(17-16)^2}{16} + \frac{(28-32)^2}{32} + \frac{(15-12)^2}{12}$

$\chi^2 = 2.625$

25. continued

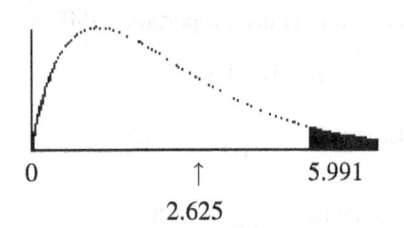

Do not reject the null hypothesis. There is not enough evidence to reject the claim that the proportions are equal.

27.

H_0: $p_1 = p_2 = p_3 = p_4 = p_5$

H_1: At least one proportion is different.

C. V. = 9.488   d. f. = 4

$E(\text{yes}) = \frac{(104)(75)}{375} = 20.8$

$E(\text{no}) = \frac{(271)(75)}{375} = 54.2$

	18	19	20	21	22
Yes	19(20.8)	18(20.8)	23(20.8)	31(20.8)	13(20.8)
No	56(54.2)	57(54.2)	52(54.2)	44(54.2)	62(54.2)

$\chi^2 = \frac{(19-20.8)^2}{20.8} + \frac{(18-20.8)^2}{20.8} + \frac{(23-20.8)^2}{20.8}$
$+ \frac{(31-20.8)^2}{20.8} + \frac{(13-20.8)^2}{20.8} + \frac{(56-54.2)^2}{54.2}$
$+ \frac{(57-54.2)^2}{54.2} + \frac{(52-54.2)^2}{54.2} + \frac{(44-54.2)^2}{54.2}$
$+ \frac{(62-54.2)^2}{54.2}$

$\chi^2 = 12.028$

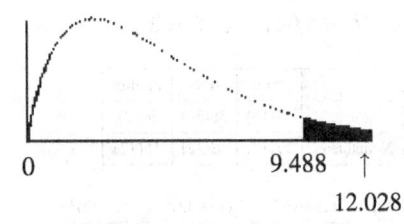

Reject the null hypothesis. There is enough evidence to conclude that the proportions are different.

29.

H_0: $p_1 = p_2 = p_3 = p_4$ (claim)

H_1: At least one proportion is different.

$\alpha = 0.05$ d. f. = 3

$E(\text{on bars}) = \frac{30(62)}{120} = 15.5$

$E(\text{not on bars}) = \frac{30(58)}{120} = 14.5$

	N	S	E	W
on	15(15.5)	18(15.5)	13(15.5)	16(15.5)
off	15(14.5)	12(14.5)	17(14.5)	14(14.5)

$\chi^2 = \frac{(15-15.5)^2}{15.5} + \frac{(18-15.5)^2}{15.5} + \frac{(13-15.5)^2}{15.5}$
$+ \frac{(16-15.5)^2}{15.5} + \frac{(15-14.5)^2}{14.5} + \frac{(12-14.5)^2}{14.5}$
$+ \frac{(17-14.5)^2}{14.5} + \frac{(14-14.5)^2}{14.5} = 1.735$

P-value > 0.10 (0.629)

(TI: P-value = 0.6291)

Do not reject the null hypothesis since P-value > 0.05. There is not enough evidence to reject the claim that the proportions are the same.

31.

H_0: $p_1 = p_2 = p_3$ (claim)

H_1: At least one proportion is different from the others.

C. V. = 7.779   d. f. = 4

Age	Inhalants	Hallucinogens	Tranquilizers
12 – 17	16(11.679)	9(13.053)	5(5.267)
18 – 25	22(23.359)	30(26.107)	8(10.534)
26 and older	13(15.962)	18(17.840)	10(7.198)

$\chi^2 = \frac{(16-11.679)^2}{11.679} + \frac{(9-13.053)^2}{13.053} + \frac{(5-5.267)^2}{5.267}$
$+ \frac{(22-23.359)^2}{23.359} + \frac{(30-26.107)^2}{26.107} + \frac{(8-10.534)^2}{10.534}$
$+ \frac{(13-15.962)^2}{15.962} + \frac{(18-17.840)^2}{17.840} + \frac{(10-7.198)^2}{7.198}$

$\chi^2 = 5.782$

Chapter 11 - Other Chi-Square Tests

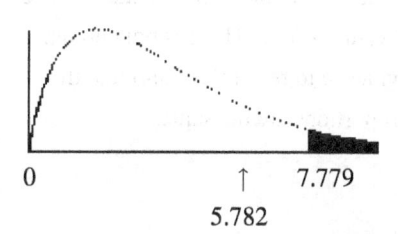

31. continued

[graph showing chi-square distribution with 5.782 marked and 7.779 marked]

0   ↑   7.779
     5.782

Do not reject the null hypothesis. There is not enough evidence to reject the claim that the proportions are equal.

33.

$$\chi^2 = \frac{(|O-E|-0.5)^2}{E} = \frac{(|12-9.61|-0.5)^2}{9.61}$$

$$+ \frac{(|15-17.39|-0.5)^2}{17.39} + \frac{(|9-11.39|-0.5)^2}{11.39}$$

$$+ \frac{(|23-20.61|-0.5)^2}{20.61}$$

$$= \frac{3.5721}{9.61} + \frac{3.5721}{17.39} + \frac{3.5721}{11.39} + \frac{3.5721}{20.61}$$

$$= 1.064$$

REVIEW EXERCISES - CHAPTER 11

1.

H_0: The distribution of trafic fatalities were as follows: used seat belt, 31.58%; did not use seat belt, 59.83%; status unknown, 8.59%. H_1: The distribution is not as stated in the null hypothesis. (claim)

C. V. = 5.991   d. f. = 2 $\alpha = 0.05$

$$\chi^2 = \sum \frac{(O-E)^2}{E} = \frac{(35-37.896)^2}{37.896} + \frac{(78-71.796)^2}{71.796}$$

$$+ \frac{(7-10.308)^2}{10.308} = 1.819$$

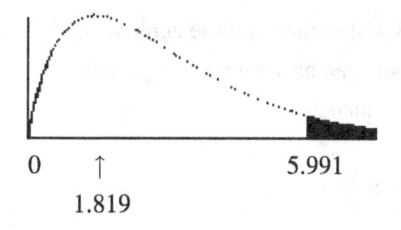

[graph showing chi-square distribution with 1.819 and 5.991 marked]

0   ↑   5.991
   1.819

1. continued

Do not reject the null hypothesis. There is not enough evidence to support the claim that the distribution differs from the one stated in the null hypothesis.

3.

H_0: The distribution of denials for gun permits is as follows: criminal history -75%, domestic violence - 11%, and other -14%. (claim)

H_1: The distribution is not the same as stated in the null hypothesis.

C. V. = 4.605   d. f. = 2

Criminal History	Domestic Violence	Other
120(150)	42(22)	38(28)

$$\chi^2 = \sum \frac{(O-E)^2}{E} = \frac{(120-150)^2}{150} + \frac{(42-22)^2}{22}$$

$$+ \frac{(38-28)^2}{28} = 27.753$$

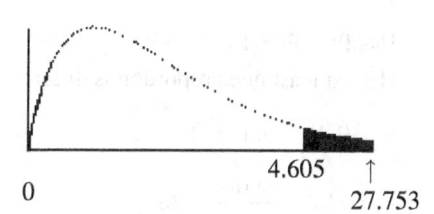

[graph showing chi-square distribution with 4.605 and 27.753 marked]

0   4.605   ↑
                      27.753

Reject the null hypothesis. There is enough evidence to reject the claim that the distribution is as stated in the null hypothesis. The distribution may vary in different geographic locations.

5.

H_0: The type of investment is independent of the age of the investor.

H_1: The type of investment is dependent upon the age of the investor. (claim)

C. V. = 9.488   d. f. = 4

5. continued

Age	Large	Small	Inter.	CD	Bond
45	20(28.18)	10(15.45)	10(15.45)	15(9.55)	45(31.36)
65	42(33.82)	24(18.55)	24(18.55)	6(11.45)	24(37.64)

$$\chi^2 = \frac{(20-28.18)^2}{28.18} + \frac{(10-15.45)^2}{15.45} + \frac{(10-15.45)^2}{15.45}$$
$$+ \frac{(15-9.55)^2}{9.55} + \frac{(45-31.36)^2}{31.36} + \frac{(42-33.82)^2}{33.82}$$
$$+ \frac{(24-18.55)^2}{18.55} + \frac{(24-18.55)^2}{18.55} + \frac{(6-11.45)^2}{11.45}$$
$$+ \frac{(24-37.64)^2}{37.64} = 27.998$$

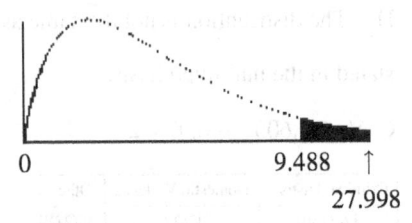

0   9.488 ↑
             27.998

Reject the null hypothesis. There is enough evidence to support the claim that the type of investment is dependent on age.

7.

H_0: $p_1 = p_2 = p_3$ (claim)
H_1: At least one proportion is different.

$\alpha = 0.01$ d. f. $= 2$

$E(\text{work}) = \frac{80(114)}{240} = 38$

$E(\text{don't work}) = \frac{80(126)}{240} = 42$

	16	17	18
work	45(38)	31(38)	38(38)
don't work	35(42)	49(42)	42(42)

$$\chi^2 = \frac{(45-38)^2}{38} + \frac{(31-38)^2}{38} + \frac{(38-38)^2}{38}$$
$$+ \frac{(35-42)^2}{42} + \frac{(49-42)^2}{42} + \frac{(42-42)^2}{42}$$

$\chi^2 = 4.912$

$0.05 < $ P-value $ < 0.10$ (0.086)

7. continued

Do not reject the null hypothesis since P-value > 0.01. There is not enough evidence to reject the claim that the proportions are the same.

9.

H_0: $p_1 = p_2 = p_3 = p_4$
H_1: At least one proportion is different from the others. (claim)

C. V. $= 7.815$ d. f. $= 3$ $\alpha = 0.05$

	Parents	Spouse	Non-family	Other
10 years ago	47(52)	38(29)	5(7)	10(12)
Now	57(52)	20(29)	9(7)	14(12)

$$\chi^2 = \frac{(47-52)^2}{52} + \frac{(38-29)^2}{29} + \frac{(5-7)^2}{7}$$
$$+ \frac{(10-12)^2}{12} + \frac{(57-52)^2}{52} + \frac{(20-29)^2}{29}$$
$$+ \frac{(9-7)^2}{7} + \frac{(14-12)^2}{12}$$

$\chi^2 = 8.357$

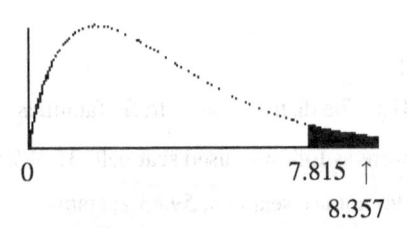

0   7.815 ↑
             8.357

Reject the null hypothesis. There is enough evidence to support the claim that at least one proportion is different from the others.

CHAPTER 11 QUIZ

1. False, it is one-tailed right.

3. False, there is little agreement between observed and expected frequencies.

5. b

7. 6

9. Right

11. H_0: The reasons why people lost their jobs are equally distributed. (claim)
H_1: The reasons why people lost their jobs are not equally distributed.

C. V. = 5.991   d. f. = 2   E = 24

$\chi^2 = \sum \frac{(O-E)^2}{E} = 2.333$

Do not reject the null hypothesis. There is not enough evidence to reject the claim that the reasons why people lost their jobs are equally distributed. The results could have been different 10 years ago since different factors of the economy existed then.

13. H_0: College students show the same preference for shopping channels as those surveyed.
H_1: College students show a different preference for shopping channels. (claim)
C. V. = 7.815   d. f. = 3
$\chi^2 = 21.789$
Reject the null hypothesis. There is enough evidence to support the claim that college students show a different preference for shopping channels.

15. H_0: Ice cream flavor is independent of the gender of the purchaser. (claim)
H_1: Ice cream flavor is dependent upon the gender of the purchaser.
C. V. = 7.815   d. f. = 3
$\chi^2 = 7.198$
Do not reject the null hypothesis. There is not enough evidence to reject the claim that ice cream flavor is independent of the gender of the purchaser.

17. H_0: The color of the pennant purchased is independent of the gender of the purchaser. (claim)
H_1: The color of the pennant purchased is dependent on the gender of the purchaser. C. V. = 4.605   d. f. = 2
$\chi^2 = 5.632$
Reject the null hypothesis. There is enough evidence to reject the claim that the color of the pennant purchased is independent of the gender of the purchaser.

19. H_0: $p_1 = p_2 = p_3$ (claim)
H_1: At least one proportion is different from the others.
C. V. = 4.605   d. f. = 2
$\chi^2 = 6.711$
Reject the null hypothesis. There is enough evidence to reject the claim that the proportions are equal. It seems that more women are undecided about their jobs. Perhaps they want better income or greater chances of advancement.

Chapter 12 - Analysis of Variance

Note: Graphs are not to scale and are intended to convey a general idea. Answers may vary due to rounding.

EXERCISE SET 12-1

1.
The analysis of variance using the F-test can be used to compare 3 or more means.

3.
The populations from which the samples were obtained must be normally distributed. The samples must be independent of each other. The variances of the populations must be equal, and the samples should be random.

5.
H_0: $\mu_1 = \mu_2 = ... = \mu_n$
H_1: At least one mean is different from the others.

7.
H_0: $\mu_1 = \mu_2 = \mu_3 = \mu_4$
H_1: At least one mean is different from the others. (claim)

C. V. = 3.01 $\alpha = 0.05$
d. f. N. = 3   d. f. D. = 24

$\overline{X}_1 = 52.286$ $s_1 = 67.243$
$\overline{X}_2 = 19.571$ $s_2 = 12.541$
$\overline{X}_3 = 35.286$ $s_3 = 16.849$
$\overline{X}_4 = 25.429$ $s_4 = 15.490$

$\overline{X}_{GM} = \frac{928}{28} = 33.143$

$s_B^2 = \frac{\sum n_i (\overline{X}_i - \overline{X}_{GM})^2}{k-1}$

$s_B^2 =$
$\frac{7(52.286-33.143)^2+7(19.571-33.143)^2+7(35.286-33.143)^2+7(25.429-33.143)^2}{3}$

$s_B^2 = 1434.421$

7. continued

$s_W^2 = \frac{\sum (n_i - 1)s_i^2}{\sum (n_i - 1)}$

$s_W^2 = \frac{6(67.243)^2+6(12.541)^2+6(16.849)^2+6(15.490)^2}{6+6+6+6}$

$s_W^2 = 1300.682$

$F = \frac{s_B^2}{s_W^2}$

$F = \frac{1434.421}{1300.682} = 1.10$

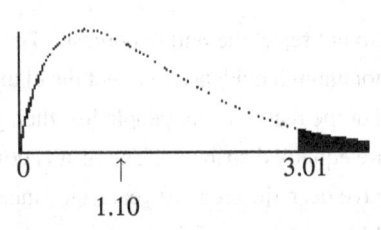

Do not reject the null hypothesis. There is not enough evidence to support the claim that at least one mean is different from the others.

9.
H_0: $\mu_1 = \mu_2 = \mu_3$
H_1: At least one mean is different from the others. (claim)

C. V. = 4.26 $\alpha = 0.05$
d. f. N. = 2   d. f. D. = 9

$\overline{X}_1 = 1.888$ $s_1 = 0.535$
$\overline{X}_2 = 2.224$ $s_2 = 0.1328$
$\overline{X}_3 = 3.525$ $s_3 = 0.1344$

$\overline{X}_{GM} = \frac{27.61}{12} = 2.301$

$s_B^2 = \frac{5(1.888-2.301)^2+5(2.224-2.301)^2+2(3.525-2.301)^2}{3-1}$

$s_B^2 = 1.9394$

$s_W^2 = \frac{4(0.535)^2+4(0.1328)^2+1(0.1344)^2}{4+4+1}$

$s_W^2 = 0.1371$

Chapter 12 - Analysis of Variance

9. continued

$F = \dfrac{1.9394}{0.1371} = 14.146$ or 14.15

(TI: $F = 14.1489$)

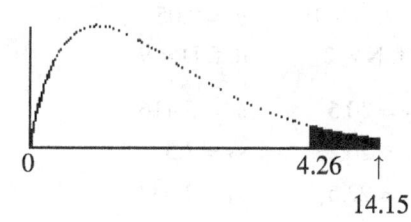

Reject the null hypothesis. There is enough evidence to conclude that at least one mean is different from the others.

11.

$H_0: \mu_1 = \mu_2 = \mu_3$

$H_1:$ At least one mean is different from the others. (claim)

C. V. = 3.89 $\alpha = 0.05$
d. f. N = 2   d. f. D = 12

$\overline{X}_1 = 32.10$ $s_1 = 16.88$
$\overline{X}_2 = 19.96$ $s_2 = 2.07$
$\overline{X}_3 = 25.98$ $s_3 = 1.65$

$\overline{X}_{GM} = \dfrac{390.2}{15} = 26.01$

$s_B^2 = 184.23$

$s_W^2 = 97.31$

$F = \dfrac{184.23}{97.31} = 1.89$

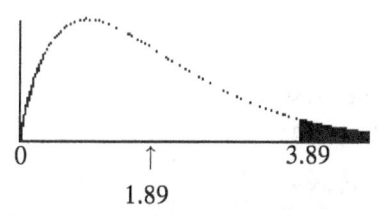

Do not reject the null hypothesis. There is not enough evidence to support the claim that at least one mean is different from the others.

13.

$H_0: \mu_1 = \mu_2 = \mu_3$

$H_1:$ At least one mean is different. (claim)

$k = 3$ $N = 18$ d.f.N. = 2   d.f.D. = 15
CV = 3.68

$\overline{X}_1 = 7$ $s_1^2 = 1.37$
$\overline{X}_2 = 8.12$ $s_2^2 = 0.64$
$\overline{X}_3 = 5.23$ $s_3^2 = 2.66$
$\overline{X}_{GM} = 6.7833$

$s_B^2 = \dfrac{6(7-6.78)^2}{2} + \dfrac{6(8.12-6.78)^2}{2}$
$\quad + \dfrac{6(5.23-6.78)^2}{2} = 12.7$

$s_W^2 = \dfrac{5(1.37)+5(0.64)+5(2.66)}{5+5+5} = 1.56$

$F = \dfrac{12.7}{1.56} = 8.14$

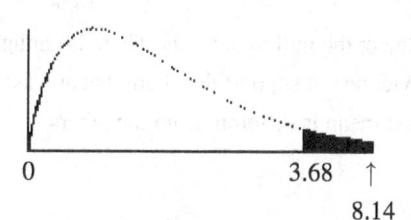

Reject the null hypothesis. There is enough evidence to support the claim that at least one mean is different.

15.

$H_0: \mu_1 = \mu_2 = \mu_3$

$H_1:$ At least one mean is different from the others. (claim)

C. V. = 2.64 $\alpha = 0.10$
d. f. N = 2   d. f. D = 17

$\overline{X}_1 = 96.33$ $s_1 = 43.80$
$\overline{X}_2 = 30$ $s_2 = 6.34$
$\overline{X}_3 = 27$ $s_3 = 15.70$

$\overline{X}_{GM} = \dfrac{983}{20} = 49.15$

Chapter 12 - Analysis of Variance

15. continued

$$s_B^2 = \frac{\sum n_i(\bar{X}_i - \bar{X}_{GM})^2}{k-1}$$

$$s_B^2 = \frac{6(96.33-49.15)^2 + 9(30-49.15)^2 + 5(27-49.15)^2}{2}$$

$$= 9554.665$$

$$s_W^2 = \frac{\sum(n_i-1)s_i^2}{\sum(n_i-1)}$$

$$= \frac{5(43.80)^2 + 8(6.34)^2 + 4(15.70)^2}{5+8+4} = 641.16$$

$$F = \frac{s_B^2}{s_W^2} = \frac{9554.665}{641.16} = 14.90$$

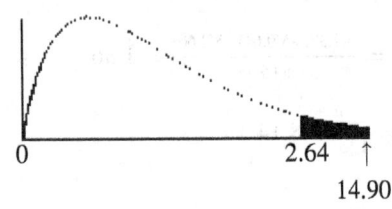

0   2.64 ↑
        14.90

Reject the null hypothesis. There is enough evidence to support the claim that at least one mean is different from the others.

17.

$H_0: \mu_1 = \mu_2 = \mu_3$
H_1: At least one mean is different from the others. (claim)

$\bar{X}_1 = 233.33$ $s_1 = 28.225$
$\bar{X}_2 = 203.125$ $s_2 = 39.364$
$\bar{X}_3 = 155.625$ $s_3 = 28.213$
$\bar{X}_{GM} = 194.091$

$$s_B^2 = \frac{21{,}729.735}{2} = 10{,}864.8675$$

$$s_W^2 = \frac{20{,}402.083}{19} = 1073.794$$

$$F = \frac{s_B^2}{s_W^2} = \frac{10{,}864.8675}{1073.794} = 10.12$$

P-value = 0.00102
Reject since P-value < 0.10. There is enough evidence to conclude that at least one mean is different from the others.

19.

$H_0: \mu_1 = \mu_2 = \mu_3$
H_1: At least one mean differs from the others. (claim)

C. V. = 3.01 $\alpha = 0.05$
d. f. N = 2   d. f. D = 9

$\bar{X}_1 = 20.5$ $s_1 = 3.416$
$\bar{X}_2 = 26.25$ $s_2 = 3.5$
$\bar{X}_3 = 22.5$ $s_3 = 2.082$

$$\bar{X}_{GM} = \frac{277}{12} = 23.083$$

$$s_B^2 = \frac{68.167}{2} = 34.083$$

$$s_W^2 = \frac{84.75}{9} = 9.417$$

$$F = \frac{34.083}{9.417} = 3.62$$

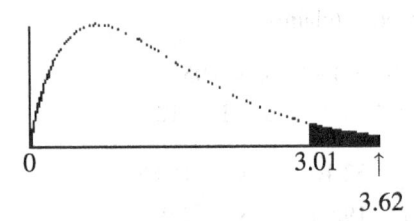

0   3.01 ↑
        3.62

Reject the null hypothesis. There is enough evidence to support the claim that at least one mean is different from the others.

EXERCISE SET 12-2

1.
The Scheffe' and Tukey tests are used.

3.
Scheffe´ Test
C. V. = 8.52

$$F_s = \frac{(\bar{X}_i - \bar{X}_j)^2}{s_W^2\left(\frac{1}{n_i} + \frac{1}{n_j}\right)}$$

For $\bar{X}_1$ vs $\bar{X}_2$

$$F_s = \frac{(1.888 - 2.224)^2}{0.13707\left(\frac{1}{5} + \frac{1}{5}\right)} = \frac{0.112896}{0.054828} = 2.10$$

Chapter 12 - Analysis of Variance

3. continued

For $\overline{X}_1$ vs $\overline{X}_3$

$$F_s = \frac{(1.888-3.525)^2}{0.13707\left(\frac{1}{5}+\frac{1}{2}\right)} = \frac{2.679769}{0.095949} = 27.93$$

For $\overline{X}_2$ vs $\overline{X}_3$

$$F_s = \frac{(2.224-3.525)^2}{0.13707\left(\frac{1}{5}+\frac{1}{2}\right)} = \frac{1.692601}{0.0959504} = 17.64$$

There is a significant difference between $\overline{X}_1$ and $\overline{X}_3$ and between $\overline{X}_2$ and $\overline{X}_3$.

5.
Tukey Test:
C. V. = 3.67

$\overline{X}_1 = 7.0$
$\overline{X}_2 = 8.12$
$\overline{X}_3 = 5.23$

$\overline{X}_1$ vs $\overline{X}_2$:
$$q = \frac{7-8.12}{\sqrt{\frac{1.56}{6}}} = -2.20$$

$\overline{X}_1$ vs $\overline{X}_3$:
$$q = \frac{7-5.23}{\sqrt{\frac{1.56}{6}}} = 3.47$$

$\overline{X}_2$ vs $\overline{X}_3$:
$$q = \frac{8.12-5.23}{\sqrt{\frac{1.56}{6}}} = 5.67$$

There is a significant difference between $\overline{X}_1$ and $\overline{X}_3$ and between $\overline{X}_2$ and $\overline{X}_3$. One reason for the difference might be that students are enrolled in cyber schools with different fees.

7.
Scheffe' Test
C. V. = 8.20

$\overline{X}_1$ vs $\overline{X}_2$:

$$F_s = \frac{(\overline{X}_i - \overline{X}_j)^2}{s_W^2\left(\frac{1}{n_i}+\frac{1}{n_j}\right)} = \frac{(850-914)^2}{9722\left(\frac{1}{4}+\frac{1}{5}\right)}$$

$F_s = 0.94$

7. continued

$\overline{X}_1$ vs $\overline{X}_3$:
$$F_s = \frac{(850-575)^2}{9722\left(\frac{1}{4}+\frac{1}{4}\right)} = 15.56$$

$\overline{X}_2$ vs $\overline{X}_3$:
$$F_s = \frac{(914-575)^2}{9722\left(\frac{1}{4}+\frac{1}{5}\right)} = 26.27$$

There is a significant difference between $\overline{X}_1$ and $\overline{X}_3$ and between $\overline{X}_2$ and $\overline{X}_3$.

9.
H_0: $\mu_1 = \mu_2 = \mu_3$
H_1: At least one mean is different from the others. (claim)

C. V. = 3.68 $\alpha = 0.05$
d. f. N = 2   d. f. D = 15

$\overline{X}_1 = 32.333$ $s_1 = 8.140$
$\overline{X}_2 = 27.833$ $s_2 = 5.529$
$\overline{X}_3 = 22.5$ $s_3 = 4.370$
$\overline{X}_{GM} = 27.556$

$$s_B^2 = \frac{290.778}{2} = 145.389$$

$$s_W^2 = \frac{579.667}{15} = 38.644$$

$$F = \frac{s_B^2}{s_W^2} = \frac{145.389}{38.644} = 3.76$$

Reject the null hypothesis. At least one mean is different from the others.

Tukey Test:
C. V. = 3.67

$$q = \frac{\overline{X}_i - \overline{X}_j}{\sqrt{\frac{s_W^2}{n}}}$$

$\overline{X}_1$ vs $\overline{X}_2$:
$$q = \frac{(32.333-27.833)}{\sqrt{\frac{38.644}{6}}} = 1.77$$

$\overline{X}_1$ vs $\overline{X}_3$:
$$q = \frac{(32.333-22.5)}{\sqrt{\frac{38.644}{6}}} = 3.87$$

9. continued

$\overline{X}_2$ vs $\overline{X}_3$:

$$q = \frac{(27.833-22.5)}{\sqrt{\frac{38.644}{6}}} = 2.10$$

There is a significant difference between $\overline{X}_1$ and $\overline{X}_3$.

11.
H_0: $\mu_1 = \mu_2 = \mu_3$
H_1: At least one mean is different from the others. (claim)

C. V. = 3.47 $\alpha = 0.05$
d. f. N. = 2   d. f. D. = 21

$\overline{X}_{GM} = 4.554$ $s_B^2 = 9.82113$

$s_W^2 = 4.93225$

$$F = \frac{9.82113}{4.93225} = 1.99$$

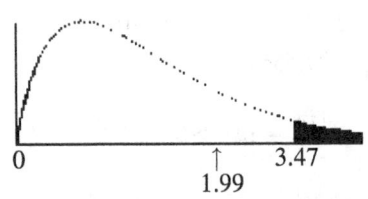

Do not reject the null hypothesis. There is not enough evidence to support the claim that at least one mean is different from the others.

13.
H_0: $\mu_1 = \mu_2 = \mu_3$
H_1: At least one mean is different. (claim)

C. V. = 3.68 $\alpha = 0.05$
d. f. N = 2   d. f. D = 15

$\overline{X}_{GM} = \frac{4666}{18} = 259.22$

$s_B^2 = \frac{39{,}374.111}{2} = 19{,}687.056$

$s_W^2 = \frac{17{,}197}{15} = 1146.467$

$$F = \frac{19{,}687.056}{1146.467} = 17.17$$

13. continued

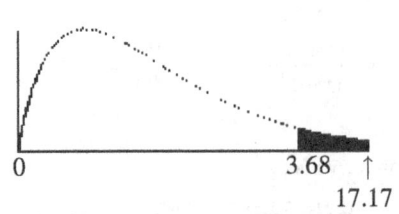

Reject the null hypothesis. There is enough evidence to support the claim that at least one mean is different.

Tukey Test:
C. V. = 3.67

$\overline{X}_1$ vs $\overline{X}_2$:

$$q = \frac{(208.17-321.17)}{\sqrt{\frac{1146.467}{6}}} = -8.17$$

$\overline{X}_1$ vs $\overline{X}_3$:

$$q = \frac{(208.17-248.33)}{\sqrt{\frac{1146.467}{6}}} = -2.91$$

$\overline{X}_2$ vs $\overline{X}_3$:

$$q = \frac{(321.17-248.33)}{\sqrt{\frac{1146.467}{6}}} = 5.27$$

There is a significant difference between $\overline{X}_1$ and $\overline{X}_2$ and between $\overline{X}_2$ and $\overline{X}_3$.

EXERCISE SET 12-3

1.
The two-way ANOVA allows the researcher to test the effects of two independent variables and a possible interaction effect. The one-way ANOVA can test the effects of one independent variable only.

3.
The mean square values are computed by dividing the sum of squares by the corresponding degrees of freedom.

Chapter 12 - Analysis of Variance

5.

a. $d.f._A = (3 - 1) = 2$ for factor A

b. $d.f._B = (2 - 1) = 1$ for factor B

c. $d.f._{AxB} = (3 - 1)(2 - 1) = 2$

d. $d.f._{within} = 3 \cdot 2(5 - 1) = 24$

7.

The two types of interactions that can occur are ordinal and disordinal.

9.

For interaction:

H_0: There is no interaction between the amount of glycerin additive and the soap concentration.

H_1: There is an interaction between the amount of glycerin additives.

For glycerin additives:

H_0: There is no difference in the means of the glycerin additives.

H_1: There is a difference in the means of the glycerin additives.

For soap concentrations:

H_0: There is no difference in the means of the soap concentrations.

H_1: There is a difference in the means of the soap concentrations.

ANOVA SUMMARY TABLE

Source	SS	d.f.	M.S.	F
Soap additive	100.00	1	100.00	5.39
Glycerin	182.25	1	182.25	9.83
Interaction	272.25	1	272.25	14.68
Within	222.50	12	18.54	
Total	777.0	15		

The critical value at $\alpha = 0.05$ with d.f. N = 1 and d.f. D = 12 is 4.75 for F_A, F_B and F_{AxB}.

9. continued

All F test values exceed the critical value, so the decision is to reject all null hypotheses. There is a significant difference at $\alpha = 0.05$ for interaction, for soap additive, and for glycerin concentration.

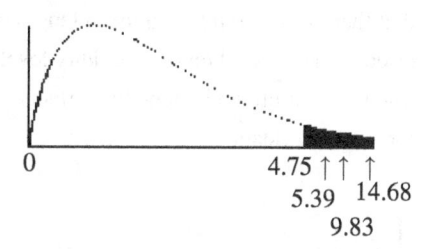

0   4.75 ↑ ↑ ↑
       5.39 14.68
          9.83

11.

For interaction:

H_0: There is no interaction effect between temperature and level of humidity.

H_1: There is an interactive effect between temperature and level of humidity.

For humidity:

H_0: There is no difference in mean length of effectiveness with respect to humidity.

H_1: There is a difference in mean length of effectiveness with respect to humidity.

For temperature:

H_0: There is no difference in mean length of effectiveness based on temperature.

H_1: There is a difference in mean length of effectiveness based on temperature.

ANOVA SUMMARY TABLE

Source	SS	d.f.	MS	F	P-value
Humidity	280.3333	1	280.3333	18.38	0.003
Temperature	3	1	3	0.197	0.669
Interaction	65.3333	1	65.3333	4.284	0.0722
Within	122	8	15.25		
Total	470.6667	11			

11. continued

The critical value at $\alpha = 0.05$ with d. f. N = 1 and d. f. D = 8 is 5.32 for F_A, F_B, and $F_{A \times B}$.

Since the only F test value that exceeds the critical value is the one for humidity, there is sufficient evidence to conclude that there is a difference in mean length of effectiveness based on the humidity level. The temperature and interaction effects are not significant.

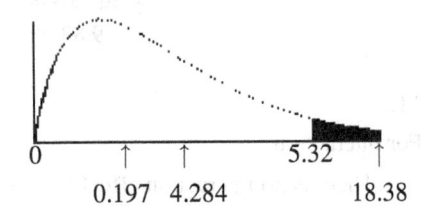

13.

For interaction:

H_0: There is no interaction effect on the durability rating between the dry additives and the solution-based additives.

H_1: There is an interaction effect on the durability rating between the dry additives and the solution-based additives.

For solution-based additive:

H_0: There is no difference in the mean durability rating with respect to the solution-based additives.

H_1: There is a difference in the mean durability rating with respect to the solution-based additives.

For dry additives:

H_0: There is no difference in the mean durability rating with respect to the dry additive.

13. continued

H_1: There is a difference in the mean durability rating with respect to the dry additive.

ANOVA SUMMARY TABLE

Source	SS	d.f.	MS	F	P-value
Solution	1.563	1	1.563	0.50	0.494
Dry	0.063	1	0.063	0.020	0.890
Interaction	1.563	1	1.563	0.50	0.494
Within	37.750	12	3.146		
Total	40.939	15			

The critical value at $\alpha = 0.05$ with d. f. N = 1 and d. f. D = 12 is 4.75. F = 0.50 for the solution-based additive and F = 0.020 for the dry additive. There is not enough evidence to conclude an effect on the mean durability based on either type of additive. For interaction, there is also not a significant interaction effect.

15.

For interaction:

H_0: There is no interaction effect between the ages of the salespersons and the products they sell on the monthly sales.

H_1: There is an interaction effect between the ages of the salespersons and the products they sell on the monthly sales.

For age:

H_0: There is no difference in the means of the monthly sales of the two age groups.

H_1: There is a difference in the means of the monthly sales of the two age groups.

For products:

H_0: There is no difference among the means of the sales for the different products.

H_1: There is a difference among the means of the sales for the different products.

Chapter 12 - Analysis of Variance

15. continued

ANOVA SUMMARY TABLE

Source	SS	d. f.	MS	F
Age	168.033	1	168.033	1.57
Product	1762.067	2	881.034	8.22
Interaction	7955.267	2	3977.634	37.09
Within	2574.000	24	107.250	
Total	12459.367	29		

At $\alpha = 0.05$, the critical values are:

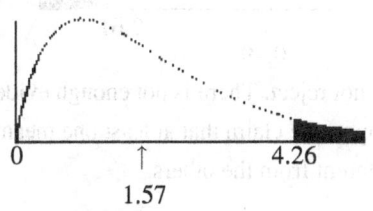

For age, d. f. N = 1, d. f. D = 24,
C. V. = 4.26

For product and interaction, d. f. N = 2,
d. f. D = 24, and C. V. = 3.40

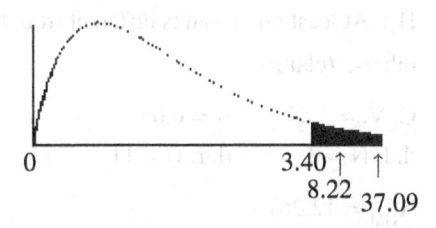

The null hypotheses for the interaction effect and for the type of product sold are rejected since the F test values exceed the critical value, 3.40.

The cell means are:

Age	Pools	Spas	Saunas
over 30	38.8	28.6	55.4
30 & under	21.2	68.6	18.8

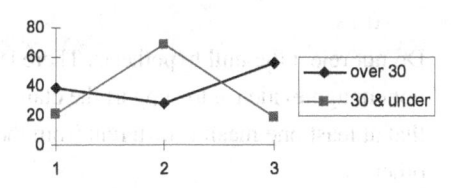

15. continued

Since the lines cross, there is a disordinal interaction hence there is an interaction effect between the age of the sales person and the type of products sold.

REVIEW EXERCISES - CHAPTER 12

1.
H_0: $\mu_1 = \mu_2 = \mu_3$ (claim)

H_1: At least one mean is different from the others.

C. V. = 5.39 $\alpha = 0.01$

d. f. N = 2   d. f. D = 33

$\bar{X}_1 = 620.5$ $s_1^2 = 5445.91$

$\bar{X}_2 = 610.17$ $s_2^2 = 22,108.7$

$\bar{X}_3 = 477.83$ $s_3^2 = 5280.33$

$\bar{X}_{GM} = \frac{20,502}{36} = 569.5$

$s_B^2 = \frac{151,890.667}{2} = 75,945.333$

$s_W^2 = \frac{361,184.333}{33} = 10,944.9798$

$F = \frac{s_B^2}{s_W^2} = \frac{75,945.333}{10,944.9798} = 6.94$

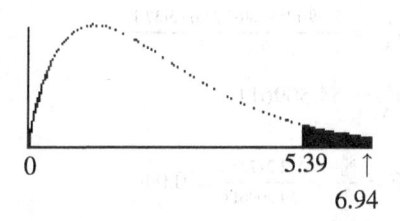

Reject. At least one mean is different.

Tukey Test   C. V. = 4.45

$\bar{X}_1$ vs $\bar{X}_2$

$q = \frac{\bar{X}_1 - \bar{X}_2}{\sqrt{\frac{s_W^2}{n}}} = \frac{620.5 - 610.17}{\sqrt{\frac{10,944.98}{12}}} = 0.34$

$\bar{X}_1$ vs $\bar{X}_3$

$q = \frac{620.5 - 477.83}{\sqrt{\frac{10,944.98}{12}}} = 4.72$

Chapter 12 - Analysis of Variance

1. continued

$\bar{X}_2$ vs $\bar{X}_3$

$q = \dfrac{610.17 - 477.83}{\sqrt{\dfrac{10{,}944.98}{12}}} = 4.38$

There is a significant difference between $\bar{X}_1$ and $\bar{X}_3$.

3.

$H_0: \mu_1 = \mu_2 = \mu_3$

H_1: At least one mean is different from the others. (claim)

C. V. = 3.55 $\alpha = 0.05$
d. f. N = 2   d. f. D = 18

$\bar{X}_1 = 29.625$ $s_1^2 = 59.125$
$\bar{X}_2 = 29$ $s_2^2 = 63.333$
$\bar{X}_3 = 28.5$ $s_3^2 = 37.1$
$\bar{X}_{GM} = 29.095$

$s_B^2 = \dfrac{\sum n_i (\bar{X}_i - \bar{X}_{GM})^2}{k-1}$

$s_B^2 = \dfrac{8(29.625 - 29.095)^2}{2} + \dfrac{7(29 - 29.095)^2}{2}$

$\quad + \dfrac{6(28.5 - 29.095)^2}{2} = 2.21726$

$s_W^2 = \dfrac{\sum (n_i - 1) s_i^2}{\sum (n_i - 1)}$

$s_W^2 = \dfrac{7(59.125) + 6(63.333) + 5(37.1)}{7 + 6 + 5}$

$s_W^2 = 54.509611$

$F = \dfrac{s_B^2}{s_W^2} = \dfrac{2.21726}{54.509611} = 0.04$

Do not reject the null hypothesis. There is not enough evidence to support the claim that at least one mean is different from the others.

5.

$H_0: \mu_1 = \mu_2 = \mu_3$

H_1: At least one mean is different. (claim)

C. V. = 2.61 $\alpha = 0.10$

5. continued

d. f. N = 2   d. f. D = 19

$\bar{X}_{GM} = 3.8591$

$s_B^2 = 1.65936$

$s_W^2 = 3.40287$

$F = \dfrac{1.65936}{3.40287} = 0.49$

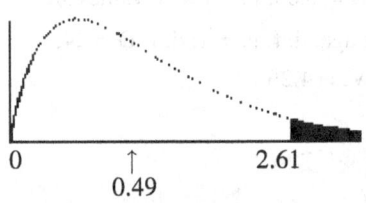

Do not reject. There is not enough evidence to support the claim that at least one mean is different from the others.

7.

$H_0: \mu_1 = \mu_2 = \mu_3 = \mu_4$

H_1: At least one mean is different from the others. (claim)

C. V. = 3.59 $\alpha = 0.05$
d. f. N = 3   d. f. D = 11

$\bar{X}_{GM} = 12.267$

$s_B^2 = 21.422$

$s_W^2 = 117.697$

$F = \dfrac{21.422}{117.697} = 0.18$

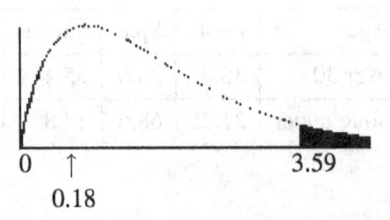

Do not reject the null hypothesis. There is not enough evidence to support the claim that at least one mean is different from the others.

Chapter 12 - Analysis of Variance

9.
H_0: There is no interaction effect between type of formula delivery system and review organization.
H_1: There is an interaction effect between type of formula delivery system and review organization.

H_0: There is no difference in mean scores based on who leads the review.
H_1: There is a difference in mean scores based on who leads the review.

H_0: There is no difference in mean scores based on who provides the formulas.
H_1: There is a difference in mean scores based on who provides the formulas.

ANOVA SUMMARY TABLE

Source	SS	d. f.	MS	F	P-value
Leaders	288.8	1	288.8	5.24	0.036
Formulas	51.2	1	51.2	0.93	0.349
Interaction	5	1	5	0.09	0.767
Within	881.2	16	55.075		
Total	1226.2	19			

At $\alpha = 0.05$ the d. f. N = 1 and the d. f. D = 16. The critical value is 4.49.

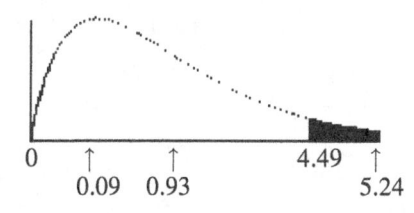

↑ 0.09   ↑ 0.93   ↑ 4.49   ↑ 5.24

There is sufficient evidence to conclude a difference in mean scores based on who leads the review.

CHAPTER 12 QUIZ

1. False, there could be a significant difference between only some of the means.

3. False, the null hypothesis should not be rejected.

5. d

7. a

9. ANOVA

11. H_0: $\mu_1 = \mu_2 = \mu_3$
H_1: At least one mean is different from the others. (claim)
C. V. = 8.02 $\alpha = 0.01$
$s_B^2 = 0.30451$ $s_W^2 = 0.00392$
$F = \frac{0.30451}{0.00392} = 77.68$

Reject the null hypothesis. There is enough evidence to support the claim that at least one mean is different from the others.

Tukey Test:
C. V. = 5.43
$\overline{X}_1 = 3.195$
$\overline{X}_2 = 3.633$
$\overline{X}_3 = 3.705$
$\overline{X}_1$ vs $\overline{X}_2$: q = −13.99
$\overline{X}_1$ vs $\overline{X}_3$: q = −16.29
$\overline{X}_2$ vs $\overline{X}_3$: q = −2.30
There is a significant difference between $\overline{X}_1$ and $\overline{X}_2$ and between $\overline{X}_1$ and $\overline{X}_3$.

13. H_0: $\mu_1 = \mu_2 = \mu_3$
H_1: At least one mean is different from the others. (claim)
C. V. = 6.93 $\alpha = 0.01$
$s_B^2 = 119.467$ $s_W^2 = 34.167$
$F = \frac{119.467}{34.167} = 3.497$

Do not reject the null hypothesis. There is not enough evidence to support the claim that at least one mean is different from the others. Writers would want to target their material to the age group of the viewers.

15. H_0: $\mu_1 = \mu_2 = \mu_3$ (claim)
H_1: At least one mean is different from the others.

C. V. = 4.46 $\alpha = 0.05$

$s_B^2 = 2114.985$ $s_W^2 = 317.958$

$F = \dfrac{2114.985}{317.958} = 6.65$

Reject the null hypothesis. There is enough evidence to support the claim that the means are not the same.

Scheffé Test:
C. V. = 8.90
For $\bar{X}_1$ vs $\bar{X}_2$, $F_S = 9.32$
For $\bar{X}_1$ vs $\bar{X}_3$, $F_S = 10.13$
For $\bar{X}_2$ vs $\bar{X}_3$, $F_S = 0.13$
There is a significant difference between $\bar{X}_1$ and $\bar{X}_2$ and between $\bar{X}_1$ and $\bar{X}_3$.

17.

a. two-way ANOVA

b. diet and exercise program

c. 2

d. H_0: There is no interaction effect between the type of exercise program and the type of diet on a person's weight loss.
H_1: There is an interaction effect between the type of exercise program and the type of diet on a person's weight loss.

H_0: There is no difference in the means of the weight losses for those in the exercise programs.
H_1: There is a difference in the means of the weight losses for those in the exercise programs.

H_0: There is no difference in the means of the weight losses for those in the diet programs.
H_1: There is a difference in the means of the weight losses for those in the diet programs.

17. continued

e. Diet: F = 21.0, significant
Exercise Program: F = 0.429, not significant
Interaction: F = 0.429, not significant

f. Reject the null hypothesis for the diets.

Chapter 13 - Nonparametric Statistics

Note: Graphs are not to scale and are intended to convey a general idea. Answers may vary due to rounding.

EXERCISE SET 13-1

1.
Non-parametric means hypotheses other than those using population parameters can be tested; distribution free means no assumptions about the population distributions have to be satisfied.

3.
Non-parametric methods have the following advantages:
a. They can be used to test population parameters when the variable is not normally distributed.
b. They can be used when data are nominal or ordinal in nature.
c. They can be used to test hypotheses other than those involving population parameters.
d. The computations are easier in some cases than the computations of the parametric counterparts.
e. They are easier to understand.
f. There are fewer assumptions that have to be met, and the assumptions are easier to verify.

5.
Distribution-free means the samples can be selected from populations that are not normally distributed.

7.
DATA 25 36   36   39 63 68 74
RANK   1 2.5 2.5   4   5   6   7

9.
DATA 2.1 6.2 11.4 12.7 18.6 20.7 22.5
RANK   1   2   3   4   5   6   7

11.
DATA 12   22   22   38 44 50
RANK   1 2.5 2.5   4   5   6

DATA 54   56   56   62 73 88
RANK   7 8.5 8.5 10 11 12

EXERCISE SET 13-2

1.
The sign test uses only + or − signs.

3.
The smaller number of + or − signs

5.
+ + + + −
+ − − − −
+ + + + +

H_0: Median = 27
H_1: Median ≠ 27 (claim)

$\alpha = 0.05$ n = 15
C. V. = 3
Test value = 5

Since 5 > 3, do not reject the null hypothesis. There is not enough evidence to support the claim that the median age is not 27 years.

7.
− + + − −
+ + − − +
− + +

H_0: Median = $35,642
H_1: Median > $35,642 (claim)

Chapter 13 - Nonparametric Statistics

7. continued

$\alpha = 0.05 \quad n = 13$
C. V. = 3   Test value = 6

Since $6 > 2$, do not reject the null hypothesis. There is not enough evidence to support the claim that the median is greater than $35,642.

9.

H_0: Median = 25   (claim)
H_1: Median > 25

C. V. = +1.65

$$z = \frac{(x - 0.5) - \left(\frac{n}{2}\right)}{\frac{\sqrt{n}}{2}} = \frac{(31 - 0.5) - \frac{50}{2}}{\frac{\sqrt{50}}{2}}$$

$$z = \frac{5.5}{3.536} = 1.56$$

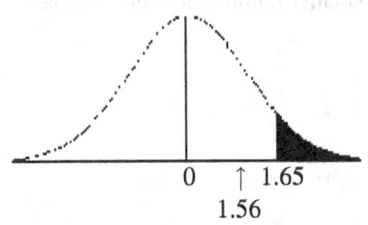

Do not reject the null hypothesis. There is not enough evidence to support the claim that more than 50% of the students favor the summer institute.

11.

H_0: Median = 150   (claim)
H_1: Median $\neq$ 150
C. V. = ± 1.96

$$z = \frac{(x + 0.5) - \left(\frac{n}{2}\right)}{\frac{\sqrt{n}}{2}} = \frac{(9 + 0.5) - \frac{35}{2}}{\frac{\sqrt{35}}{2}}$$

$z = -2.70$

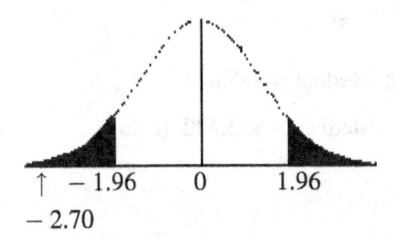

11. continued

Reject. There is enough evidence to reject the claim that the median number of faculty members is 150.

13.

H_0: Median = 49   (claim)
H_1: Median $\neq$ 49

$$z = \frac{(x + 0.5) - \left(\frac{n}{2}\right)}{\frac{\sqrt{n}}{2}} = \frac{(36 + 0.5) - \frac{98}{2}}{\frac{\sqrt{98}}{2}}$$

$z = -2.53$
P-value = 0.0114

Reject. There is enough evidence to reject the claim that 50% of the students are against extending the school year.

15.

A	B	C	D	E	F	G	H
+	+	−	+	+	−	+	+

H_0: The number of sessions will not be reduced.
H_1: The number of sessions will be reduced. (claim)

$\alpha = 0.05 \quad n = 8$

C. V. = 1   Test value = 2

Since $2 > 1$, do not reject the null hypothesis. There is not enough evidence to support the claim that the number of sessions was reduced.

17.

A	B	C	D	E	F	G	H	I	J
−	+	+	+	+	+	−	−	+	+

H_0: The number of soft drinks will not change.
H_1: The number of soft drinks will change. (claim)

Chapter 13 - Nonparametric Statistics

17. continued

$\alpha = 0.10$ n = 10

C. V. = 1   Test value = 3

Since 3 > 1, do not reject the null hypothesis. There is not enough evidence to support the claim that the number of soft drinks was reduced

19.

1	2	3	4	5	6	7	8	9	10
+	+	+	+	+	+	−	+	+	−

H_0: The number of viewers is the same as last year. (claim)

H_1: The number of viewers is not the same as last year.

$\alpha = 0.01$ n = 10

C. V. = 0   Test value = 2

Since 2 > 0, do not reject the null hypothesis. There is not enough evidence to reject the claim that the number of viewers is the same as last year.

21.

3, 4, 6, 9, 12, 15, 15, 16, 18, 22, 25, 30

At $\alpha = 0.05$, the value from Table J with n = 12 is 2; hence, count in 3 numbers from each end to get $6 \leq MD \leq 22$.

23.

4.2, 4.5, 4.7, 4.8, 5.1, 5.2, 5.6, 6.3, 7.1, 7.2, 7.8, 8.2, 9.3, 9.3, 9.5, 9.6

At $\alpha = 0.02$, the value from Table J with n = 16 is 2; hence, count 3 numbers from each end to get $4.7 \leq MD \leq 9.3$.

25.

12, 14, 14, 15, 16, 17, 18, 19, 19, 21, 23, 25, 27, 32, 33, 35, 39, 41, 42, 47

25. continued

At $\alpha = 0.05$, the value from Table J with n = 20 is 5; hence, count in 6 numbers from each end to get $17 \leq MD \leq 33$.

EXERCISE SET 13-3

1.

The sample sizes n_1 and n_2 must be greater than or equal to 10.

3.

H_0: There is no difference in the speed skating times of the students at the two universities. (claim)

H_1: There is a difference in the speed skating times of the students at the two universities.

C. V. = ± 1.96

1:35	1:38	1:39	1:40	1:42
UB	UB	UA	UA	UA
1	2	3	4	5

1:48	1:48	1:51	1:58	2:00
UB	UB	UA	UA	UB
6.5	6.5	8	9	10

2:01	2:03	2:05	2:06	2:10
UA	UA	UA	UB	UB
11	12	13	14	15

2:14	2:15	2:15	2:20	2:27
UB	UA	UB	UA	UB
16	17.5	17.5	19	20

R = 101.5

$\mu_R = \dfrac{10(10 + 10 + 1)}{2} = 105$

$\sigma_R = \sqrt{\dfrac{10 \cdot 10(10 + 10 + 1)}{12}} = 13.23$

$Z = \dfrac{101.5 - 105}{13.23} = -0.26$

Chapter 13 - Nonparametric Statistics

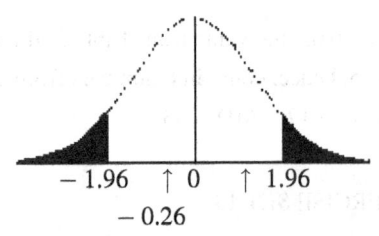

3. continued

$-1.96 \quad \uparrow 0 \quad \uparrow 1.96$
-0.26

Do not reject the null hypothesis. There is not enough evidence to reject the claim that there is no difference in the times.

5.
H_0: There is no difference in the number of credits transferred.
H_1: There is a difference in the number of credits transferred. (claim)

C. V. = $\pm 1.96 \quad \alpha = 0.05$

35	37	42	45	46	48	58	58	59
1	2	3	4	5	6	7.5	7.5	9
C	S	C	S	S	C	S	S	C

60	61	62	63	63	64	64
10	11	12	13.5	13.5	15.5	15.5
C	C	C	C	S	C	S

65	66	68	71
17	18	19	20
C	S	S	S

$R = 98$ (for community college)

$\mu_R = \dfrac{n_1(n_1 + n_2 + 1)}{2}$

$\mu_R = \dfrac{10(10 + 10 + 1)}{2} = \dfrac{10(21)}{2} = \dfrac{210}{2} = 105$

$\sigma_R = \sqrt{\dfrac{n_1 \cdot n_2(n_1 + n_2 + 1)}{12}}$

$\sigma_R = \sqrt{\dfrac{10 \cdot 10(10 + 10 + 1)}{12}} = \sqrt{\dfrac{(10)(10)(21)}{12}}$

$\sigma_R = \sqrt{175} = 13.23$

5. continued

$Z = \dfrac{R - \mu_R}{\sigma_R} = \dfrac{98 - 105}{13.23} = -0.53$

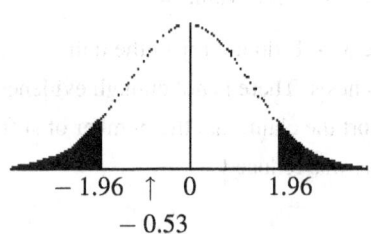

$-1.96 \quad \uparrow 0 \quad 1.96$
-0.53

Do not reject the null hypothesis. There is not enough evidence to support the claim that there is a difference in the number of credits transferred.

7.
H_0: There is no difference between the stopping distances of the two types of automobiles. (claim)
H_1: There is a difference between the stopping distances of the two types of automobiles.

C. V. = ± 1.65

186	187	188	188	190	192	193
1	2	3.5	3.5	5	6	7
M	M	M	M	M	M	C

194	195	196	198	200	203	204
8	9	10	11	12	13	14
M	M	C	C	C	M	C

206	211	212	214	218	297
15	16	17	18	19	20
C	C	C	M	C	C

$R = 69$

$\mu_R = \dfrac{10(10 + 10 + 1)}{2} = 105$

$\sigma_R = \sqrt{\dfrac{10 \cdot 10(10 + 10 + 1)}{12}} = 13.23$

$z = \dfrac{69 - 105}{13.23} = -2.72$

Chapter 13 - Nonparametric Statistics

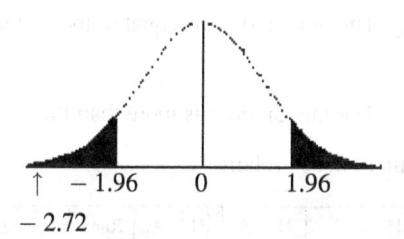

7. continued

↑   −1.96   0   1.96
−2.72

Reject the null hypothesis. There is enough evidence to reject the claim that there is no difference in the stopping distances of the two types of automobiles.

9.
H_0: There is no difference in the number of hunting accidents in the two regions.
H_1: There is a difference in the number of hunting accidents in the two regions. (claim)

C. V. = ±1.96

2	3	5	5	6	7	8	8	9	10	11
1	2	3.5	3.5	5	6	7.5	7.5	9	10	12
E	E	E	E	E	E	E	W	W	W	W

11	11	13	13	14	15	17	17	21
12	12	14.5	14.5	16	17	18.5	18.5	20
W	E	E	W	E	W	W	W	W

R = 71

$$\mu_R = \frac{n_1(n_1+n_2+1)}{2} = \frac{10(10+10+1)}{2} = 105$$

$$\sigma_R = \sqrt{\frac{n_1 \cdot n_2(n_1+n_2+1)}{12}}$$

$$\sigma_R = \sqrt{\frac{10 \cdot 10(10+10+1)}{12}} = 13.23$$

$$Z = \frac{R - \mu_R}{\sigma_R} = \frac{71-105}{13.23} = -2.57$$

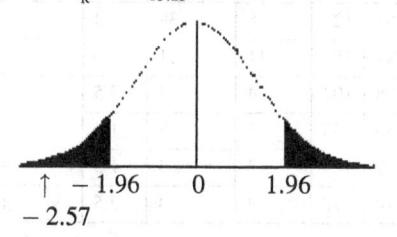

↑   −1.96   0   1.96
−2.57

9. continued

Reject the null hypothesis. There is enough evidence to support the claim that there is a difference in the number of accidents in the two areas. The number of accidents may be related to the number of hunters in the two areas.

11.
H_0: There is no difference in job satisfaction.
H_1: There is a difference in job satisfaction (claim).

C. V. = ±1.65

51	57	57	59	62	63	65	68	69
1	2.5	2.5	4	5	6	7	8	9
M	U	M	M	M	M	M	U	U

71	73	76	78	79	80	83
10	11	12	13	14	15	16
M	U	U	U	U	M	M

84	85	86	87	92	94	95	96	98	99
17	18	19	20	21	22	23	24	25	26
U	U	M	U	M	U	M	M	U	U

R = 153.5

$$\mu_R = \frac{n_1(n_1+n_2+1)}{2} = \frac{13(13+13+1)}{2}$$

$$\mu_R = \frac{13(27)}{2} = 175.5$$

$$\sigma_R = \sqrt{\frac{n_1 \cdot n_2(n_1+n_2+1)}{12}}$$

$$\sigma_R = \sqrt{\frac{13 \cdot 13(13+13+1)}{12}} = 19.5$$

$$Z = \frac{R - \mu_R}{\sigma_R} = \frac{153.5-175.5}{19.5} = -1.13$$

Chapter 13 - Nonparametric Statistics

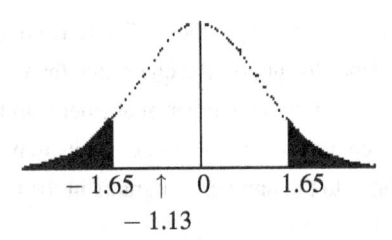

11. continued

−1.65 ↑ 0 1.65
−1.13

Do not reject the null hypothesis. There is not enough evidence to support the claim that there is a difference in job satisfaction between the two groups.

EXERCISE SET 13-4

1.
The t-test for dependent samples

3.

B	A	B − A	|B − A|	Rank	Signed Rank
106	112	− 6	6	3	− 3
85	84	1	1	1	1
117	105	12	12	6.5	6.5
163	167	− 4	4	2	− 2
154	142	12	12	6.5	6.5
106	113	− 7	7	4	− 4
152	143	9	9	5	5

Sum of the − ranks: $(-3) + (-2) + (-4) = -9$.
Sum of the + ranks: $1 + 6.5 + 6.5 + 5 = 19$
$w_s = 9$

5.
C. V. = 16 $w_s = 13$
Since $13 \leq 16$, reject the null hypothesis.

7.
C. V. = 60 $w_s = 65$
Since $65 > 60$, do not reject the null hypothesis.

9.
H_0: The human dose is equal to the animal dose.
H_1: The human dose is more than the animal dose. (claim)

H	A	H − A	|H − A|	Rank	Signed Rank
0.67	0.13	0.54	0.54	7	+ 7
0.64	0.18	0.46	0.46	6	+ 6
1.20	0.42	0.78	0.78	8	+ 8
0.51	0.25	0.26	0.26	4	+ 4
0.87	0.57	0.30	0.30	5	+ 5
0.74	0.57	0.17	0.17	3	+ 3
0.50	0.49	0.01	0.01	1	+ 1
1.22	1.28	− 0.06	0.06	2	− 2

Sum of the − ranks: − 2
Sum of the + ranks: + 34

$n = 8$ C. V. = 6
$w_s = |-2| = 2$
Since $2 < 6$, reject the null hypothesis. There is enough evidence to support the claim that the human dose costs more than the animal dose. One reason is that some people might not be inclined to pay a lot of money for their pets' medication.

11.
H_0: The amount spent on lottery tickets does not change.
H_1: The amount spent on lottery tickets is reduced. (claim)

$n = 8$ $\alpha = 0.05$ C. V. = 6

B	A	B − A	|B − A|	Rank	Signed Rank
86	72	14	14	7	+ 7
150	143	7	7	5	+ 5
161	123	38	38	8	+ 8
197	186	11	11	6	+ 6
98	102	− 4	4	3.5	− 3.5
56	53	3	3	1.5	+ 1.5
122	125	− 3	3	1.5	− 1.5
76	72	4	4	3.5	+ 3.5

Chapter 13 - Nonparametric Statistics

11. continued

Sum of the − ranks: − 5
Sum of the + ranks: + 31
$w_s = 5$

Since $5 \le 6$, reject the null hypothesis. There is enough evidence to support the claim that the workshop reduced the amount the participants spent on lottery tickets.

13.

H_0: The prices of prescription drugs in the United States are equal to the prices in Canada.

H_1: The drugs sold in Canada are cheaper. (claim)

$n = 10$ $\alpha = 0.05$ C. V. = 11

U.S.	C	US − C	|US − C|	Rank	Signed Rank
3.31	1.47	1.84	1.84	8	+ 8
2.27	1.07	1.20	1.20	4.5	+ 4.5
2.54	1.34	1.20	1.20	4.5	+ 4.5
3.13	1.34	1.79	1.79	7	+ 7
23.40	21.44	1.94	1.94	10	+ 10
3.16	1.47	1.69	1.69	6	+ 6
1.98	1.07	0.91	0.91	3	+ 3
5.27	3.39	1.88	1.88	9	+ 9
1.96	2.22	− 0.26	0.26	2	− 2
1.11	1.13	− 0.02	0.02	1	− 1

Sum of the + ranks:
$8 + 4.5 + 4.5 + 7 + 10 + 6 + 3 + 9 = 52$
Sum of the − ranks: $(-2) + (-1) = -3$
$w_s = |-3| = 3$

Since $3 < 11$, reject the null hypothesis. There is enough evidence to support the claim that the drugs are less expensive in Canada.

EXERCISE SET 13-5

1.

H_0: There is no difference in the results of the questionnaires among the three groups.

1. continued

H_1: There is a difference in the results of the questionnaires among the three groups. (claim)

C. V. = 5.991

A	Rank	B	Rank	C	Rank
22	9	18	6.5	16	3
25	12.5	24	11	17	4.5
27	15.5	25	12.5	19	8
26	14	27	15.5	23	10
33	21	29	17	18	6.5
35	22	31	19.5	31	19.5
30	18	17	4.5	15	1.5
36	23.5	15	1.5	36	23.5
$R_1 =$	135.5	$R_2 =$	88	$R_3 =$	76.5

$$H = \frac{12}{N(N+1)} \left(\frac{R_1^2}{n_1} + \frac{R_2^2}{n_2} + \frac{R_3^2}{n_3} \right) - 3(N+1)$$

$$H = \frac{12}{24(24+1)} \left(\frac{135.5^2}{8} + \frac{88^2}{8} + \frac{76.5^2}{8} \right) - 3(24+1) = 4.891$$

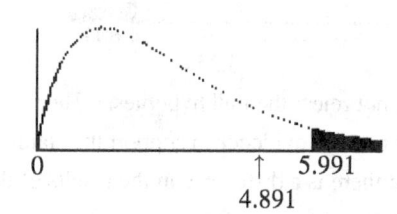

Do not reject the null hypothesis. There is not enough evidence to support the claim that there is a difference in the results of the questionnaire.

3.

H_0: There is no difference in the scores on the questionnaire.

H_1: There is a difference in the scores on the questionnaire. (claim)

C. V. = 4.605

3. continued

NM	Rank	M	Rank	D	Rank
37	14.5	40	19	38	16
39	17.5	36	13	35	12
32	8	32	8	21	2
31	5.5	33	10.5	19	1
37	14.5	39	17.5	31	5.5
32	8	33	10.5	24	3
		30	4		

$R_1 = 68$ $R_2 = 82.5$ $R_3 = 39.5$

$$H = \frac{12}{N(N+1)} \left(\frac{R_1^2}{n_1} + \frac{R_2^2}{n_2} + \frac{R_3^2}{n_3} \right) - 3(N+1)$$

$$H = \frac{12}{19(19+1)} \left(\frac{68^2}{6} + \frac{82.5^2}{7} + \frac{39.5^2}{6} \right) - 3(19+1)$$

$H = 3.254$

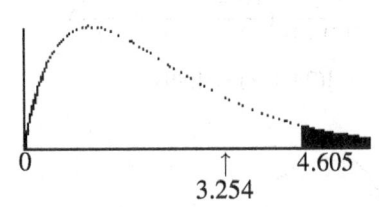

0 ↑ 4.605
  3.254

Do not reject the null hypothesis. There is not enough evidence to support the claim that there is a difference in the results of the questionnaire.

5.

H_0: There is no difference in the sugar content of the three different types of candy bars.

H_1: There is a difference in the sugar content of the three different types of candy bars. (claim)

C. V. = 5.991

5. continued

A	Rank	B	Rank	C	Rank
7.6	1	9.2	3	18.6	16
9.3	4	11.1	9	16.2	13
8.4	2	12.3	11	15.4	12
11.3	10	10.1	6	18	15
10.2	7.5	10.2	7.5	17.3	14
9.8	5				

$R_1 = 29.5$ $R_2 = 36.5$ $R_3 = 70$

$$H = \frac{12}{N(N+1)} \left(\frac{R_1^2}{n_1} + \frac{R_2^2}{n_2} + \frac{R_3^2}{n_3} + \frac{R_4^2}{n_4} \right) - 3(N+1)$$

$$= \frac{12}{16(16+1)} \left(\frac{29.5^2}{6} + \frac{36.5^2}{5} + \frac{70^2}{2} \right) - 3(16+1)$$

$H = 10.389$

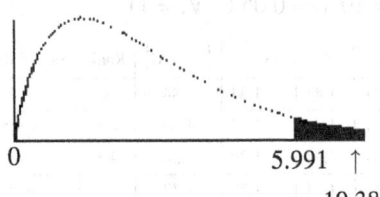

0 5.991 ↑
         10.389

Reject the null hypothesis. There is enough evidence to support the claim that the sugar content of three of candy bars is different.

7.

H_0: There is no difference in spending between regions.

H_1: There is a difference in spending between regions.

C. V. = 5.991

E	Rank	M	Rank	W	Rank
6701	3	9854	15	7584	10
6708	4	8414	11	5474	1
9186	12	7279	7	6622	2
6786	5	7311	8	9673	14
9261	13	6947	6	7353	9

$R_1 = 37$ $R_2 = 47$ $R_3 = 36$

Chapter 13 - Nonparametric Statistics

7. continued

$$H = \frac{12}{N(N+1)} \left(\frac{R_1^2}{n_1} + \frac{R_2^2}{n_2} + \frac{R_3^2}{n_3} + \frac{R_4^2}{n_4} \right) - 3(N+1)$$

$$H = \frac{12}{15(15+1)} \left(\frac{37^2}{5} + \frac{47^2}{5} + \frac{36^2}{5} \right) - 3(15+1)$$

$$H = 0.74$$

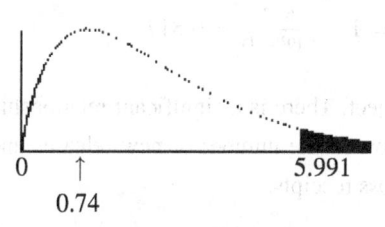

0 ↑ 5.991
0.74

Do not reject the null hypothesis. There is not enough evidence to conclude that there is a difference in spending.

9.

H_0: There is no difference in the number of crimes in the 5 precincts.

H_1: There is a difference in the number of crimes in the 5 precincts. (claim)

C. V. = 13.277

1	Rank	2	Rank	3	Rank
105	24	87	13	74	7.5
108	25	86	12	83	11
99	22	91	16	78	9
97	20	93	18	74	7.5
92	17	82	10	60	5
$R_1=$	108	$R_2=$	69	$R_3=$	40

4	Rank	5	Rank
56	3	103	23
43	1	98	21
52	2	94	19
58	4	89	15
62	6	88	14
$R_4=$	16	$R_5=$	92

$$H = \frac{12}{N(N+1)} \left(\frac{R_1^2}{n_1} + \frac{R_2^2}{n_2} + \frac{R_3^2}{n_3} + \frac{R_4^2}{n_4} + \frac{R_5^2}{n_5} \right) - 3(N+1)$$

9. continued

$$H = \frac{12}{25(25+1)} \left(\frac{108^2}{5} + \frac{69^2}{5} + \frac{40^2}{5} + \frac{16^2}{5} + \frac{92^2}{5} \right) - 3(25+1)$$

$$H = 20.753$$

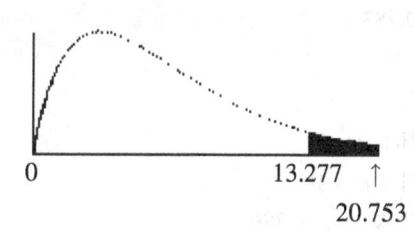

0   13.277 ↑
              20.753

Reject the null hypothesis. There is enough evidence to support the claim that there is a difference in the number of crimes for the precincts.

11.

H_0: There is no difference in speeds.

H_1: There is a difference in speeds. (claim)

C. V. = 5.991

Pred.	Rank	Deer	Rank	Dom.	Rank
70	15	50	12.5	47.5	11
50	12.5	35	5.5	39.35	7
43	10	32	4	35	5.5
42	9	30	2.5	30	2.5
40	8	61	14	11	1
$R_1=$	54.5	$R_2=$	38.5	$R_3=$	27

$$H = \frac{12}{N(N+1)} \left(\frac{R_1^2}{n_1} + \frac{R_2^2}{n_2} + \frac{R_3^2}{n_3} \right) - 3(N+1)$$

$$H = \frac{12}{15(15+1)} \left(\frac{54.5^2}{5} + \frac{38.5^2}{5} + \frac{27^2}{5} \right) - 3(15+1)$$

$$H = 3.815$$

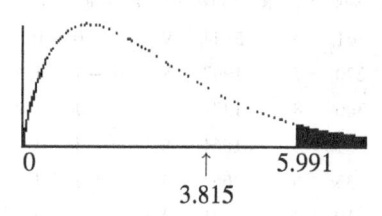

0   ↑   5.991
      3.815

Do not reject. There is not enough evidence to conclude there is a difference in speeds.

Chapter 13 - Nonparametric Statistics

EXERCISE SET 13-6

1.
0.392

3.
0.783

5.
H_0: $\rho = 0$
H_1: $\rho \neq 0$
C. V. = ± 0.786

Grade 4	R_1	Grade 8	R_2	$R_1 - R_2$	d^2
90	7	81	7	0	0
84	4	75	4	0	0
80	3	66	2	1	1
87	5	76	5	0	0
88	6	80	6	0	0
77	1	59	1	0	0
79	2	74	3	−1	1
				$\sum d^2 =$	2

$r_s = 1 - \frac{6 \cdot \sum d^2}{n(n^2-1)} = 1 - \frac{6 \cdot 2}{7(7^2-1)}$

$r_s = 0.964$

Reject the null hypothesis. There is a significant relationship between the mathematics achievement scores for the two grades.

7.
H_0: $\rho = 0$
H_1: $\rho \neq 0$
C. V. = ± 0.700

Releases	R_1	Receipts	R_2	$R_1 - R_2$	d^2
361	9	3844	9	0	0
270	7	1967	8	−1	1
306	8	1371	7	1	1
22	5	1064	6	−1	1
35	6	667	5	1	1
10	2	241	4	−2	4
8	1	188	3	−2	4

7. continued

Releases	R_1	Receipts	R_2	$R_1 - R_2$	d^2
12	3	154	2	1	1
21	4	125	1	3	9
				$\sum d^2 =$	22

$r_s = 1 - \frac{6 \cdot \sum d^2}{n(n^2-1)}$

$r_s = 1 - \frac{6 \cdot 22}{9(9^2-1)} = 0.817$

Reject. There is a significant relationship between the number of new releases and gross receipts.

9.
H_0: $\rho = 0$
H_1: $\rho \neq 0$
C. V. = ± 0.738

Calories	R_1	Cholesterol	R_2	$R_1 - R_2$	d^2
580	7.5	205	3	4.5	20.25
580	7.5	225	6	1.5	2.25
270	1	285	8	−7	49
470	6	270	7	−1	1
420	4	185	1.5	2.5	6.25
415	3	215	4	−1	1
330	2	185	1.5	0.5	0.25
430	5	220	5	0	0
				$\sum d^2 =$	80

$r_s = 1 - \frac{6 \cdot 80}{8(8^2-1)} = 1 - \frac{480}{504} = 0.048$

Do not reject the null hypothesis. There is not enough evidence to say that a significant relationship exists between calories and cholesterol amounts in fast-food sandwiches.

11.
H_0: $\rho = 0$
H_1: $\rho \neq 0$
C. V. = ± 0.786

11. continued

Instructor	R_1	Student	R_2	$R_1 - R_2$	d^2
1	1	2	2	-1	1
4	4	6	6	-2	4
6	6	7	7	-1	1
7	7	5	5	2	4
5	5	4	4	1	1
2	2	3	3	-1	1
3	3	1	1	2	4
				$\sum d^2 =$	16

$$r_s = 1 - \frac{6\sum d^2}{n(n^2-1)} = 1 - \frac{6 \cdot 16}{7(7^2-1)} = 0.714$$

Do not reject the null hypothesis. There is not enough evidence to say that there is a relationship in the rankings of the textbook between the instructors and the students.

13.

H_0: $\rho = 0$

H_1: $\rho \neq 0$

C. V. $= \pm 0.900$

Students	R_1	Cost	R_2	$R_1 - R_2$	d^2
10	3	7200	2	1	1
6	1	9393	5	-4	16
17	5	7385	3	2	4
8	2	4500	1	1	1
11	4	8203	4	0	0
				$\sum d^2 =$	22

$$r_s = 1 - \frac{6\sum d^2}{n(n^2-1)}$$

$$r_s = 1 - \frac{6 \cdot 22}{5(5^2-1)} = -0.10$$

Do not reject the null hypothesis. There is no significant relationship between the number of cyber school students and the cost per pupil. In this case, the cost per pupil is different in each district.

15.

H_0: The occurrance of cavities is random.

H_1: The null hypothesis is not true.

The median of the data set is two. Using A = above and B = below, the runs (going across) are shown:

B AA B AAA B A BB AAAA B A B
A B A B A B AAA B A BB

There are 21 runs. The expected number of runs is between 10 and 22. Therefore, the null hypothesis should not be rejected. The number of cavities occurs at random.

17.

H_0: The type of admissions occur at random. (claim)

H_1: The null hypothesis is not true.

SS F S F S FF S F SS FFF S FF SSS FFFF S
FFF S F SS F SS FF SS

There are 23 runs and this value is between 13 and 27. Hence, do not reject the null hypothesis. The admissions occur at random.

19.

H_0: The ups and downs in the stock market occur at random. (claim)

H_1: The ups and downs in the stock market do not occur at random.

UUUU DD UUU D UUUUU DDD U D

There are 8 runs and since this is between 5 and 15, the null hypothesis is not rejected. The ups and downs in the stock market occur at random.

21.

H_0: The number of absences of employees occurs at random. (claim)

H_1: The null hypothesis is not true.

Chapter 13 - Nonparametric Statistics

21. continued

The median of the data is 12. Using A = above and B = below, the runs are shown as follows:

A B AAAAAAA BBBBBBBB AAAAAA BBBB

There are 6 runs. The expected number of runs is between 9 and 21, hence the null hypothesis is rejected since 6 is not between 9 and 21. The number of absences do not occur at random.

23.

H_0: The number of on-demand movie rentals occurs at random. (claim)

H_1: The null hypothesis is not true.

The median of the data is 6.5. Using A = above and B = below, the runs are shown as follows:

BB AAA BB A B A BB AAAA BBB A

There are 10 runs. The expected number of runs is between 6 and 16, hence the null hypothesis is not rejected since 10 is between 6 and 16. The number of movie rentals occur at random.

25.

H_0: The gender of the patients at a medical center occurs at random. (claim) H_1: The null hypothesis is not true.

C. V. $= \pm 1.96$

$n_1 = 28$ (men) and $n_2 = 22$ (women)

FF MMMMM FFF MMMMMM F MM FFFF MMM F M F MMMMMM F MM F M FF M FFFFFF M

There are $G = 20$ runs. The test statistic is:

$z = \frac{G - \mu_G}{\sigma_G}$

$\mu_G = \frac{2(28)(22)}{28+22} + 1 = 25.64$

25. continued

$\sigma_G = \sqrt{\frac{2(28)(22)[2(28)(22)-28-22]}{(28+22)^2(28+22-1)}} = 3.448$

$z = \frac{20-25.64}{3.448} = -1.636$ or -1.64

Do not reject the null hypothesis. There is not enough evidence to reject the claim that the sequence is random.

27.

H_0: The patients who were treated for an accident or illness occur at random. (claim)

H_1: The null hypothesis is not true.

C. V. $= \pm 1.96$

$n_1 = 24$ (illness) and $n_2 = 36$ (accident)

I A I AAAAAA I A II AA II AAAA I A I AAA II AA II AA I A I A I A I AA II AAA II A I AA II AAA

There are $G = 34$ runs. The test statistic is:

$z = \frac{G - \mu_G}{\sigma_G}$

$\mu_G = \frac{2(24)(36)}{24+36} + 1 = 29.8$

$\sigma_G = \sqrt{\frac{2(24)(36)[2(24)(36)-24-36]}{(24+36)^2(24+36-1)}} = 3.684$

$z = \frac{34-29.8}{3.684} = 1.14$

Do not reject the null hypothesis. There is not enough evidence to reject the claim that the sequence occurs at random.

29.

$r = \frac{\pm 1.96}{\sqrt{50-1}} = \pm 0.28$

31.

$r = \frac{\pm 2.33}{\sqrt{35-1}} = \pm 0.400$

33.

$r = \frac{\pm 2.58}{\sqrt{40-1}} = \pm 0.413$

Chapter 13 - Nonparametric Statistics

REVIEW EXERCISES - CHAPTER 13

1.
H_0: median = $9.00 (claim)
H_1: median ≠ $9.00

C. V. = ± 1.96 $\alpha = 0.05$

$$Z = \frac{(9+0.5) - \left(\frac{30}{2}\right)}{\sqrt{\frac{30}{2}}} = -2.01$$

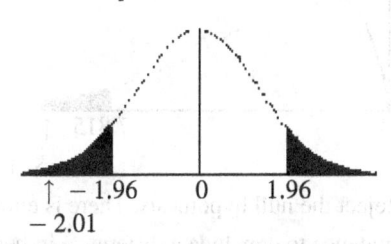

↑ −1.96   0   1.96
−2.01

Reject. There is enough evidence to reject the claim that the median price is $9.00.

3.
H_0: There is no difference in prices.
H_1: There is a difference in prices. (claim)

$R^+ = 7, R^- = 1$
C. V. = 0
Test value = 1

Do not reject the null hypothesis. There is not enough evidence to conclude a difference in prices. Comments: Examine what affects the result of this test.

5.
H_0: There is no difference in the amount of hours worked.
H_1: There is a difference in the amount of hours worked.

C. V. = ± 1.645 at $\alpha = 0.10$

12	15	17	18	18	19	19	20	20
1	2	3	4.5	4.5	6.5	6.5	8.5	8.5
A	A	A	A	L	A	L	A	L

5. continued

21	22	22	24	24	25	25	26
10	11.5	11.5	13.5	13.5	15.5	15.5	17
L	A	L	A	L	L	A	L

28	30	31	35
18	19	20	21
L	A	L	L

R = 85

$$\mu_R = \frac{n_1(n_1 + n_2 + 1)}{2} = \frac{10(10 + 11 + 1)}{2} = 110$$

$$\sigma_R = \sqrt{\frac{n_1 n_2(n_1 + n_2 + 1)}{12}}$$

$$\sigma_R = \sqrt{\frac{10 \cdot 11(10 + 11 + 1)}{12}} = 14.201$$

$$Z = \frac{R - \mu_R}{\sigma_R} = \frac{85 - 110}{14.201} = -1.76$$

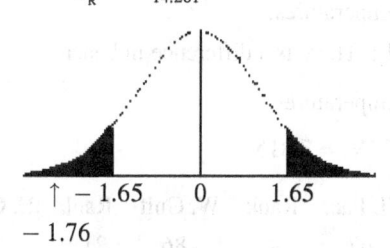

↑ −1.65   0   1.65
−1.76

Reject the null hypothesis. There is enough evidence to conclude a difference in the hours worked.

At $\alpha = 0.05$, C. V. = ± 1.96. The decision would be to not reject the null hypothesis.

7.
H_0: There is no difference in the amount spent.
H_1: There is a difference in the amount spent. (claim)

Chapter 13 - Nonparametric Statistics

7. continued

B	A	B − A	|B − A|	Rank	Signed Rank
7	6	1	1	1	1
5.5	10	− 4.5	4.5	7	− 7
4.5	7	− 2.5	2.5	6	− 6
10	12	− 2	2	4	− 4
6.75	8.5	− 1.75	1.75	2	− 2
5	7	− 2	2	4	− 4
6	8	− 2	2	4	− 4

Sum of the + ranks: 1
Sum of the − ranks: − 27
$w_s = 1$ C. V. = 2 $\alpha = 0.05$ n = 7

Reject the null hypothesis. There is enough evidence to conclude that there is a difference in amount spent at the 0.05 level of significance.

9.

H_0: There is no difference in beach temperatures.

H_1: There is a difference in beach temperatures.

C. V. = 7.815

Cl. Pac.	Rank	W. Gulf	Rank	E. Gulf	Rank
67	4	86	21	87	25
68	5	86	21	87	25
66	3	84	13	86	21
69	6	85	16.5	86	21
63	2	79	8	85	16.5
62	1	85	16.5	84	13
			.5	85	16.5
$R_1 =$	21	$R_2 =$	96	$R_3 =$	138

S. Atl.	Rank
76	7
81	10
82	11
84	13
80	9
86	21
87	25
$R_4 =$	96

9. continued

$$H = \frac{12}{N(N+1)}\left(\frac{R_1^2}{n_1} + \frac{R_2^2}{n_2} + \frac{R_3^2}{n_3}\right) - 3(N+1)$$

$$H = \frac{12}{26(26+1)}\left(\frac{21^2}{6} + \frac{96^2}{6} + \frac{138^2}{7} + \frac{96^2}{7}\right) - 3(26+1)$$

H = 15.524

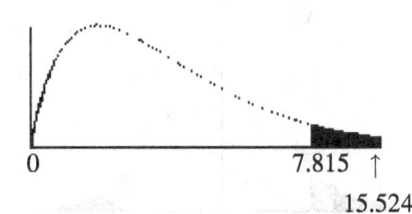

Reject the null hypothesis. There is enough evidence to conclude a difference in beach temperatures.

Without the Southern Pacific:

C. V. = 5.991 at $\alpha = 0.05$

$$H = \frac{12}{20(20+1)}\left(\frac{60^2}{6} + \frac{96^2}{7} + \frac{54^2}{7}\right) - 3(20+1)$$

H = 3.661

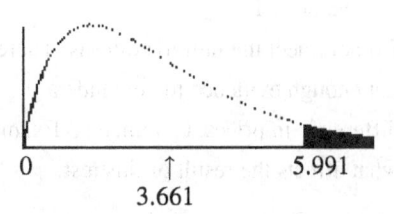

Do not reject the null hypothesis. There is not enough evidence to conclude a difference in the temperatures. The conclusion is not the same without the Southern Pacific temperatures.

11.

H_0: $\rho = 0$
H_1: $\rho \neq 0$
C. V. = ± 0.786

Chapter 13 - Nonparametric Statistics

11. continued

Pages	R_1	Sources	R_2	$R_1 - R_2$	d^2
15	1	10	1	0	0
25	4	18	4.5	-0.5	0.25
23	3	18	4.5	-1.5	2.25
30	6	15	3	3	9
18	2	13	2	0	0
28	5	23	7	-2	4
35	7	20	6	1	1
				$\sum d^2 =$	16.5

$r_s = 1 - \dfrac{6\sum d^2}{n(n^2-1)} = 1 - \dfrac{6(16.5)}{7(48)} = 0.7054$

Do not reject the null hypothesis. There is not a significant relationship between the number of pages and the number of references.

13.
H_0: The grades of students who finish the exam occur at random. (claim)
H_1: The null hypothesis is not true.

The median grade is 73. Using A = above and B = below, the runs are:

AAAA B AAAA BBBB AAAAA BB A BBBBBBB

Since there are 8 runs and this does not fall between 9 and 21, the null hypothesis is rejected. The grades do not occur at random.

CHAPTER 13 QUIZ

1. False

3. True

5. a

7. d

9. Non-parametric

11. Sign

13.
H_0: Median = \$230,500
H_1: Median $\neq$ \$230,500 (claim)
There are three $-$ signs.
Do not reject since 3 is greater than the critical value of 2. There is not enough evidence to support the claim that the median is not \$230,500.

15.
H_0: There will be no change in the weight of the turkeys after the special diet.
H_1: The turkeys will weigh more after the special diet. (claim)
There is one $+$ sign. Reject the null hypothesis since the critical value is 0. There is enough evidence to support the claim that the turkeys gained weight on the special diet.

17.
H_0: The distributions are the same.
H_1: The distributions are different. (claim)
C. V. $= \pm 1.65$
$z = -0.144$
Do not reject the null hypothesis. There is not enough evidence to support the claim that the distributions are different.

19.
H_0: There is no difference in the amounts of sodium in the three sandwiches.
H_1: There is a difference in the amounts of sodium in the sandwiches. (claim)
H = 11.795
C. V. = 5.991
Reject the null hypothesis. There is enough evidence to support the claim that there is a difference in the amounts of sodium in the three sandwiches.

Chapter 13 - Nonparametric Statistics

21.
H_0: $\rho = 0$
H_1: $\rho \neq 0$
C. V. $= \pm 0.600$
$r_s = 0.633$
Reject the null hypothesis. There is a significant relationship between the drug prices.

23.
H_0: The gender of babies occurs at random.
H_1: The null hypothesis is false.
$\alpha = 0.05$ C. V. $= 8, 19$
There are 10 runs, which is between 8 and 19. Do not reject the null hypothesis. There is not enough evidence to reject the null hypothesis that the gender occurs at random.

25.
H_0: The numbers occur at random
H_1: The null hypothesis is false.
$\alpha = 0.05$ C. V. $= 9, 21$
The median number is 538.
There are 20 runs and since this is between 9 and 21, the null hypothesis is not rejected. There is not enough evidence to reject the null hypothesis that the numbers occur at random.

Chapter 14 - Sampling and Simulation

EXERCISE SET 14-1

1.
Random, systematic, stratified, cluster.

3.
A sample must be randomly selected.

5.
Talking to people on the street, calling people on the phone, and asking one's friends are three incorrect ways of obtaining a sample.

7.
Random sampling has the advantage that each unit of the population has an equal chance of being selected. One disadvantage is that the units of the population must be numbered, and if the population is large this could be somewhat time consuming.

9.
An advantage of stratified sampling is that it ensures representation for the groups used in stratification; however, it is virtually impossible to stratify the population so that all groups could be represented.

11.
Answers will vary.

13.
Answers will vary.

15.
Answers will vary.

17.
Answers will vary.

19.
Answers will vary.

21.
Sampling or selection bias occurs when some subjects are more likely to be included in a study than others.

23.
Nonresponse bias occurs when subjects who do not respond to a survey question would answer the question differently than the subjects who responded to the survey question.

25.
Answers will vary. Response or interview bias occurs when the subject does not give his or her true opinion and gives an opinion that he or she feels is politically correct.

27.
Volunteer bias occurs when people volunteer to participate in a study because they are interested in the study or survey.

EXERCISE SET 14-2

1.
Flaw - biased; it's confusing.

3.
Flaw - the question is too broad.

5.
Flaw - confusing wording. The question could be worded: "How many hours did you study for this exam?"

7.
This question has confusing wording. Change the question to read: "If a plane were to crash on the border of New York and New Jersey, where should the victims be buried?"

Chapter 14 - Sampling and Simulation

9.
Flaw - the word "vaguely" is too general.

11.
The word *family* could mean different things to the respondent, for example, in cases of separated families.

13.
The word regularly is vague.

15.
A person might not know of the situation four years ago.

17.
This question assumes the subject feels texting while driving is bad. Not all people will agree with this.

19.
Here the question limits the response to "repeated" tours. Subjects might not be in favor of any tour.

21.
Answers will vary.

EXERCISE SET 14-3

1.
Simulation involves setting up probability experiments that mimic the behavior of real life events.

3.
John Van Neumann and Stanislaw Ulam.

5.
The steps are:
1. List all possible outcomes.
2. Determine the probability of each outcome.

5. continued
3. Set up a correspondence between the outcomes and the random numbers.
4. Conduct the experiment using random numbers.
5. Repeat the experiment and tally the outcomes.
6. Compute any statistics and state the conclusions.

7.
When the repetitions increase there is a higher probability that the simulation will yield more precise answers.

9.
Use three-digit random numbers; numbers 001 through 681 mean that the mother is in the labor force.

11.
Select 100 two-digit random numbers. Numbers 00 through 34 mean the household has at least one set with permium cable service. Numbers 35 to 99 mean the household does not have the service.

13.
Let an odd number represent heads and an even number represent tails. Then each person selects a digit at random.

15.
Answers will vary.

17.
Answers will vary.

19.
Answers will vary.

Chapter 14 - Sampling and Simulation

21.
Answers will vary.

23.
Answers will vary.

25.
Answers will vary.

REVIEW EXERCISES - CHAPTER 15

1.
Answers will vary.

3.
Answers will vary.

5.
Answers will vary.

7.
Answers will vary.

9.
Flaw - asking a biased question. Change the question to read: "Have you ever driven through a red light?"

11.
Flaw - asking a double-barreled question. Change the question to read: "Do you think all automobiles should have heavy-duty bumpers?"

13.
Use one-digit random numbers 1 through 4 to represent a strikeout and 5 through 9 and 0 to represent anything other than a strikeout.

15.
The first person selects a two-digit random number. Any two-digit random number that has a 7, 8, 9, or 0 is ignored, and another random number is selected. Player 1 selects a one-digit random number; any random number that is not 1 through 6 is ignored, and another one is selected.

17.
Answers will vary.

19.
Answers will vary.

21.
Answers will vary.

CHAPTER 14 QUIZ

1. True

3. False, only random numbers generated by a random number table are random.

5. a

7. c

9. Biased

11.
Answers will vary.

13.
Answers will vary.

15.
Use two-digit random numbers: 01 through 45 constitute a win. Any other two-digit number means the player loses.

Chapter 14 - Sampling and Simulation

17.
Use two-digit random numbers 01 through 10 to represent the 10 cards in hearts. The random numbers 11 through 20 represent the 10 cards in diamonds. The random numbers 21 through 30 represent the 10 spades, and 31 through 40 represent the 10 clubs. Any number over 40 is ignored.

19.
Use two-digit random numbers. The first digit represents the first player, and the second digit represents the second player. If both numbers are odd or even, player 1 wins. If a digit is odd and the other digit is even, player 2 wins.

21.
Answers will vary.

23.
Answers will vary.

25.
Here regularly is vague.

27.
What is meant by readable?

29.
Some respondents might not know much about herbal medicine.